POP-UP GEOMETRY

Anyone browsing at the stationery store will see an incredible array of pop-up cards available for any occasion. The workings of pop-up cards and pop-up books can be remarkably intricate. Behind such designs lies beautiful geometry involving the intersection of circles, cones, and spheres, the movements of linkages, and other constructions. The geometry can be modelled by algebraic equations, whose solutions explain the dynamics. For example, several pop-up motions rely on the intersection of three spheres, a computation made every second for GPS location. Connecting the motions of the card structures with the algebra and geometry reveals abstract mathematics performing tangible calculations. Beginning with the nephroid in the 19th-century, the mathematics of pop-up design is now at the frontiers of rigid origami and algorithmic computational complexity. All topics are accessible to those familiar with high-school mathematics; no calculus required. Explanations are supplemented by 140+ figures and 20 animations.

Joseph O'Rourke is Olin Professor of Computer Science and Professor of Mathematics at Smith College. His research is in discrete and computational geometry, developing algorithms for geometric computations. He has won several awards, including a Guggenheim Fellowship in 1987 and the NSF Director's Award for Distinguished Teaching Scholars in 2001. He was named an ACM Fellow in 2012. He has published more than 165 papers, more than 30 of which were coauthored with undergraduates. He has taught folding and unfolding to students in grade school, middle school, high school, college and graduate school, and to teachers at all educational levels, as well as to researchers in mathematics and computer science. This is his seventh book.

Pop-Up Geometry

The Mathematics
Behind Pop-Up Cards

Joseph O'Rourke
Smith College, Massachusetts

CAMBRIDGE
UNIVERSITY PRESS

University Printing House, Cambridge CB2 8BS, United Kingdom

One Liberty Plaza, 20th Floor, New York, NY 10006, USA

477 Williamstown Road, Port Melbourne, VIC 3207, Australia

314–321, 3rd Floor, Plot 3, Splendor Forum, Jasola District Centre,
New Delhi – 110025, India

103 Penang Road, #05–06/07, Visioncrest Commercial, Singapore 238467

Cambridge University Press is part of the University of Cambridge.

It furthers the University's mission by disseminating knowledge in the pursuit of
education, learning, and research at the highest international levels of excellence.

www.cambridge.org
Information on this title: www.cambridge.org/9781009098403
DOI: 10.1017/9781009093095

First published 2022

A catalogue record for this publication is available from the British Library.

ISBN 978-1-009-09840-3 Hardback
ISBN 978-1-009-09626-3 Paperback

Contents

Preface

Bertrand Russell said: "Mathematics rightly viewed possesses not only truth but supreme beauty." It is, alas, not always possible to "rightly view" mathematics from the perspective of a student. Often mathematics is presented as a series of pointless (and difficult) exercises. However, seeing mathematics applied to tangible, physical objects in motion can reveal a glimpse of that "supreme beauty." It can be uplifting to open a pop-up card and see a crease in cardstock sweep out a cone, while a particular corner tracks a circle formed by the intersection of two moving spheres. One can almost hallucinate these geometric structures guiding the intricate dynamics of the pop-up structures. This is my goal: to enable the reader to see and appreciate the mathematics underlying pop-up design.

Pop-up books and cards have been around since the eighteenth century, and recently have seen a surge in popularity through the elaborate designs of pop-up masters such as Matthew Reinhart and Robert Sabuda. This book will not help you achieve design mastery, and understanding the mathematics behind pop-ups is not even necessary to become a proficient "paper engineer." But it is satisfying to understand the mechanisms behind pop-up constructions, which gives one an appreciation of the achievements of the designers. And, most importantly for my goals, one can learn fascinating mathematics through the study of pop-up designs.

The mathematics can be surprisingly intricate. It has led to at least one Ph.D. thesis (del Rosario Ruiz, 2015)[1] and several technical academic papers. But we will only need high-school mathematics: algebra and geometry, some limited trigonometry, and no calculus. The final two chapters explore algorithms, but we make no assumption of previous exposure to aspects of computer science.

Seeing the mathematics applied to real, physical structures can be illuminating: the equations are no longer pointless exercises, but are aimed at explaining visible dynamics. Although the mathematical prerequisites are minimal, much of the reasoning is in 3D and likely novel to readers, as 3D topics are not emphasized in standard curricula. We even prove half-a-dozen theorems, uncommon in high-school instruction outside of two-column geometry proofs.

For these reasons, more than a dozen boxed explanations are sprinkled throughout the text at points where the mathematics might be new or long

---

[1]This is a citation of a references listed in the bibliography (p. 123).

forgotten. In addition, a list of the symbols used is included at the end of the book (p. 121).

The text includes over 40 exercises, which are marked as *Practice*, *Understanding*, or *Challenge*. A Practice exercise might be a simple calculation, whereas an Understanding exercise generally requires a thorough grasp of the preceding material. Challenge exercises either go beyond the text or might involve a substantial investment of time. We encourage the reader to read all the exercises, give them as much thought as inclinations and circumstances allow, and then turn to Chapter 8 at the end of the book where solutions to all exercises are provided.

All pop-up card dynamics are driven by the opening of the card. Understanding the dynamics is most easily achieved either by manipulating a physical model or via an animation. All the templates in the book are available through the author's website,[2] and links are provided there to more than 25 animated GIFs (O'Rourke, 2021).

We now offer a brief summary of each chapter.

Summary. The first four chapters, with some exceptions in Chapter 2, describe constructions built from just a single piece of cardstock, cut and creased in particular ways, without separate glued-in attachments. It is remarkable the variety of effects that can be achieved under these restrictions. Chapter 1 concentrates on establishing the mathematical vocabulary and conventions we follow throughout and on applying these to parallel cuts in the cardstock. A typical application is to pop-up letters, say, "10 YEARS OLD!" In Chapter 2 we turn to the versatile V-fold and show how its motion in 3D can be understood as the intersection of three spheres, whose equations we derive and solve. We also explain the challenges of converting the "horizontal" rotary motion of the card opening to either a "vertical" rotary motion or a "flat" rotary motion. All this scare-quoted notation will of course be explained. Chapter 3 explores a simple design based on cutting parallel chords of a circle centered on the cardline, which opens to an elegant shape we call the Knight's visor. The equations describing this shape connect to beautiful nineteenth-century mathematics and—surprisingly—to caustics, shadow-shapes formed by light passing through a glass of water. In contrast to the Knight's visor, whose final, static shape is the goal, the pop-up spinner in Chapter 4 is all dynamics — amazing dynamics. The engine driving the spinner is a theorem from an undergraduate thesis on protein folding (Benbernou, 2006), which we prove in a simplified form.

The last three chapters progress to pasting-in various structures, necessarily collapsible structures to allow the card to close. Chapter 5 focuses on popping up convex polyhedra—the cube, an octahedron, etc.—and the rich mathematics surrounding polyhedra. Unlike the classical mathematics uncovered in the Knight's visor construction, the mathematics relevant to collapsing polyhedra is quite contemporary. We even quote unsolved problems (pp. 66, 99) under active investigation.

---

[2] cs.smith.edu/~jorourke/PopUps/

The final two chapters discuss algorithms: Chapter 6 describes an algorithm that can output instructions for a pop-up of a wide class of polyhedra, wider than the convex polyhedra considered in Chapter 5, but not as wide as "all polyhedra," which is an as-yet unsolved problem. The final short chapter touches on the way computer scientists measure the difficulty of a problem, its "computational complexity." Perhaps it is no surprise that deciding whether a particular pop-up structure can collapse flat is technically "intractable," i.e., very hard!

Acknowledgements

I benefited from inspired assistance from students who worked with me on aspects of pop-up design, including Nadia Benbernou, Rachel Darling, Elizabeth Freeman, Kelsey Hammond, Serena Hansraj, Stephanie Jakus, Eindra Kyi, Qaiomei Li, Molly Miller, Duc Nuygen, Eleanor O'Rourke, Gail Parsloe, Emmely Rogers, Yujun Shen, Ana Spasova, Claire Tepesch, Maria Vespa, Lucy Wang, and Faith Weller.

I thank my Cambridge editor, Kaitlin Leach, for expert guidance throughout the process. My research was partially funded by NSF Grant DUE-0123154.

1

Parallel Folds

We start with *parallel folds*, the simplest and easiest pop-up technique to understand. Despite the simplicity of parallel folds, one can reach beautiful and intricate designs with just this one construction technique, repeated. But we'll keep it simple, making our goal pop-up letters, perhaps the most common pop-up card design.

The mathematics behind parallel folds is also simple, which will provide an opportunity to set conventions and notation that will be used throughout. We will aim toward describing the 3D motion of card opening by tracking various key points on the card front and back traveling along circles in space.

1.1   Card Notation

Here we establish the basic notation used throughout the book, which is important because the rigor of mathematics relies on clear definitions and symbolic notation.[1]

A card is composed of two identical rectangles, the back B and the front F, joined and hinged along the card centerline L. (The centerline is also called the gutter or the spine.) The front and back are both rigid, usually made of stiff cardstock. We view the back as fixed to a tabletop; only the front moves. See Fig. 1.1. In general we will use uppercase letters for "big" things and lowercase letters for "small" things, for example, edges or specific points. And we will follow the mathematics tradition of using (lowercase) Greek letters for angles: α, β, θ, etc. In particular, we will reserve θ (theta) to represent the card angle, the angle along the centerline between the front and the back. At $\theta = 0°$ the card is closed, at $\theta = 90°$ it is half-open, and at $\theta = 180°$ the card is fully opened flat. Often, the card is intended to only open to $90°$, when F becomes a backdrop to the popped-up structure. This will be the case for most of the constructions in this chapter.

---

[1] Symbols are gathered in a table on p. 121.

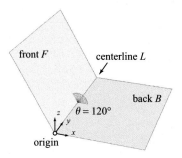

Figure 1.1 Basic card notation.

In geometry, the angle in 3D between two planes, such as θ just described, is known as a *dihedral angle*; see Box 1.1, which relies on vectors, described in Box 1.2.

Our overall goal is to describe the geometry of the motions of the pop-up structures as the card is opened or closed. Often we reach the geometry through algebraic equations. The equations will employ variables representing points in a 3D Cartesian coordinate system, called *Cartesian* because it was introduced by René Descartes. We will consistently use x and y to represent placement in the xy-plane of the card back B, and use z as the third coordinate representing the perpendicular height above B. We place the origin of the coordinate system somewhere along the centerline L, either at the bottom end or the midpoint of L, with x increasing horizontally along the width of the back and y increasing vertically along the centerline.

Box 1.1 Dihedral Angle

In geometry, the angle in 3D between two planes is known as a *dihedral angle*, from the Greek *di-* (two) and *hedra* (faces). A *face* is a region of a plane. Ultimately angles are measured in two dimensions, not three, between what are known as *vectors*: see Box 1.2.

In Fig. 1.2, F_1 and F_2 are faces sharing an edge e. The dihedral angle δ at e can be measured by a protractor perpendicular to edge e; so

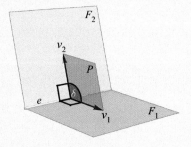

Figure 1.2 The dihedral angle δ between faces F_1 and F_2.

the protractor lies in a plane P orthogonal to e. (The term *orthogonal* is a synonym for "perpendicular.") The vectors v_1 and v_2 lie along the intersection of plane P with the planes containing the faces F_1 and F_2, respectively. The angle δ measures how much v_2 needs to be rotated about e to align with v_1. In this chapter, e is the card centerline L and δ is the card angle θ.

Box 1.2 Vectors

Informally, a *vector* can be viewed as a directed line segment. Thus it has both a length and a direction. The two endpoints are called the *tail* and the *head*, with the head marked by an arrowhead. A vector differs from a directed line segment in that its tail is not fixed to one point of the plane or space. Rather it should be imagined to be placed anywhere, as in Fig. 1.3.

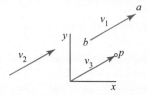

Figure 1.3 Vectors v_1, v_2, v_3 are the same vector $a - b$.

Often it is convenient to view a point p, which has specific coordinates with respect to an origin, as a vector with its tail at the origin and its head at p. Given any two points a and b, the difference in their coordinates, $a - b$, is a vector, placed with the head at a and the tail at b. Here if $a = (a_x, a_y)$ and $b = (b_x, b_y)$, then

$$a - b = (a_x - b_x, \, a_y - b_y) .$$

We will make cuts and creases in the card (and in attachments to the card). Creases in the card can be either *valley* or *mountain* folds, terminology from origami with the obvious interpretation. All creases can be viewed as hinges between the flat, planar pieces to either side. Those flat pieces are often called *faces*, which are rigid except where hinged along the edges.

1.2   Rhombus Card

Our first pop-up is a *rhombus*, a four-sided figure (a quadrilateral) all of whose edge lengths are equal. Fig. 1.4 shows the construction from a rectangle of cardstock. After folding the cardstock in half along centerline L, the two

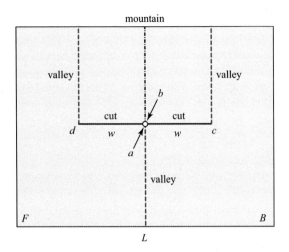

Figure 1.4 Rhombus cuts and folds, producing Fig. 1.5.

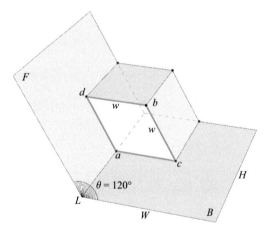

Figure 1.5 Rhombus card from the template in Fig. 1.4.

length-w cuts can be achieved by one scissors cut of length w perpendicular to L. Then the mountain and valley creases shown pop out the rhombus, as illustrated in Fig. 1.5. With several points labeled, the *rim* (green in Fig. 1.5) is a 2D rhombus $acbd$ lying in a plane orthogonal to the centerline L. The sides of the rhombus have length w. We'll use the notation ac to represent the line segment from point a to point c, and $|ac| = w$ to indicate its length (Box 1.3.)

Box 1.3 Segment Length $|ab|$

Throughout this book, we use lowercase letters to represent points, in either 2D or 3D. We often need to refer to the length of a segment

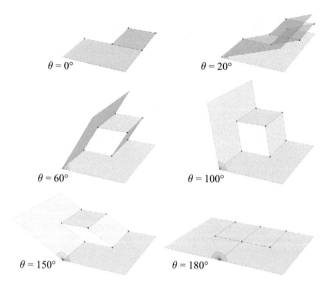

$\theta = 0°$ $\theta = 20°$

$\theta = 60°$ $\theta = 100°$

$\theta = 150°$ $\theta = 180°$

Figure 1.6 Snapshots of the rhombus card. Animation available (O'Rourke, 2021).

between two points. If a and b are two points, we use $|ab|$ to mean the length of the line segment from a to b (which of course is the same as $|ba|$, the length from b to a). In terms of 3D coordinates, with $a = (a_x, a_y, a_z)$ and $b = (b_x, b_y, b_z)$,

$$|ab| = \sqrt{(a_x - b_x)^2 + (a_y - b_y)^2 + (a_z - b_z)^2} .$$

Often this is expressed as $|a - b|$, because the difference of the point coordinates is a vector from b to a, and $|ab|$ is the length of this vector, as explained in Box 1.2. Other sources notate this as $\|a - b\|$, where $\|$ indicates the *norm* of the vector.

The cut segment creates two sides, which then become the four sides of the rhombus rim. Note that in the flattened template, points a and b have the same location—they are *co-located*—and become opposite corners of the rhombus. Fig. 1.6 shows snapshots of the card at increasing values of θ from 0°—completely closed—to 180°—fully opened, when it matches the template in Fig. 1.4.

This construction is called a *parallel fold* because the two valley creases and the one mountain crease are parallel (and parallel to the card centerline). For aesthetic reasons, often an extra layer of cardstock is pasted behind the front and back faces of the card, so that the gap formed by popping up the rhombus has a backdrop, rather than the hole shown in Fig. 1.6.

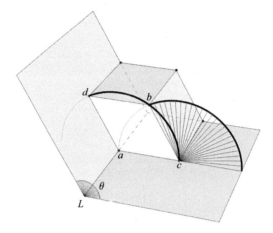

Figure 1.7 Points b and d follow circular arcs centered on c and a, respectively.

1.3 Circular Arc Equations

As is evident in Fig. 1.7, point b at the corner of the rhombus must remain the fixed distance $|cb| = w$ from point c throughout the opening motion, and therefore must ride on a circle of radius w centered on c. Similarly, point d remains distance $|ad| = w$ from a, and so follows a circle centered on a. In preparation of more complicated motions in later chapters, we now describe these motions via algebraic equations.

The equation of a radius-w circle in the xy-plane, centered on the origin, is

$$x^2 + y^2 = w^2 . \tag{1.1}$$

The circles followed by points b and d lie instead in the xz-plane, a plane orthogonal to the y-axis along the centerline L. Let's focus on point d revolving around a. One might think that the equation for d's position is

$$x^2 + z^2 = w^2 , \tag{1.2}$$

simply substituting z for y and otherwise not mentioning y. But because that equation does not constrain y, it describes circles for all values of y. So Eq. 1.2 describes an infinite cylinder centered on the y-axis (and in a 3D coordinate system, Eq. 1.1 denotes an infinite cylinder centered on the z-axis). To describe the circle that d follows, we need two equations, including one pinning down the precise y-coordinate for point a.

Let us say that the card back B has width W and height H (uppercase distinguishing from lowercase w and h). Then point a has coordinates $a = (0, \frac{H}{2}, 0)$, and d is constrained by these two equations "simultaneously":

$$x^2 + z^2 = w^2 ,$$
$$y = \frac{H}{2} .$$

We will frequently need to "solve" equations simultaneously, but here there is nothing to solve: we are just specifying that both equations constrain x, y, z at the same time.

Similarly, point b follows the circle arc centered on $c = (w, \frac{H}{2}, 0)$ defined by these two equations, where x is shifted by w:

$$(x - w)^2 + z^2 = w^2 \, ,$$

$$y = \tfrac{H}{2} \, .$$

Lastly we presage *parametric equations*, which were used to draw the spokes in Fig. 1.7 and which will play a significant role in later chapters (Box 2.2). Parametric equations rely on a *parameter*, which in our situation is the card angle θ. The equations we just derived describe the full circle for point b "all at once." Parametric equations pinpoint the location of b for a specific value of θ:

$$x = w + w \cos \theta \, ,$$
$$y = \tfrac{H}{2} \, , \qquad\qquad (1.3)$$
$$z = w \sin \theta \, .$$

For example, if $\theta = 90°$, then $\cos\theta$ and $\sin\theta$ are 0 and 1, respectively, so $(x, y, z) = (w, \frac{H}{2}, w)$, which places b directly above $c = (w, \frac{H}{2}, 0)$. (For more on cosine and sine, see Box 1.4.)

Box 1.4 Trigonometry

We will use trigonometry sparingly and never need more than the basic relationships illustrated in Fig. 1.8, always measuring angles counter-clockwise. Note that not only is $\tan\theta$ the ratio of the altitude to the base of the triangle, but it is the length of the tangent from p to the x-axis. This latter relationship helps to understand why, as θ approaches $90°$, $\tan\theta$ grows without bound—approaches infinity.

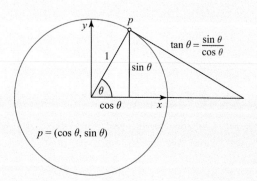

Figure 1.8 Basic trigonometry: unit-radius circle.

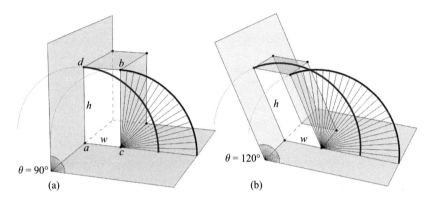

Figure 1.9 (a) Asymmetric cut: $h > w$. (b) Two parallel asymmetric cuts.

1.4   Parallelograms

Two simple modifications of the rhombus construction turn it into a useful pop-up structure. First, rather than cut a length w to each side of the centerline L, an asymmetric cut of w on the back B and h on the front F leads to a pop-up parallelogram, $w \times h$ at card angle $\theta = 90°$. See Fig. 1.9(a).

Second, two identical parallel cuts, symmetric or asymmetric about L, result in a centered pop-up, rather than the parallelogram extending to the card's top edge. See Fig. 1.9(b). With $h > w$ as in the figure, the pop-up can serve as a face into which letters can be carved. When $h < w$, the pop-up often serves as a platform on which other structures are built.

Exercise 1.1 Practice: Within Card Profile

With the card width W and height H, which combinations of the pop-up parallelogram dimensions w and h ensure that no portion of the pop-up "sticks out" when the card is fully closed at $\theta = 0°$? (See Fig. 1.6.)

Exercise 1.2 Understanding: Arcs Intersection

In Fig. 1.9(b), what are the coordinates of the point p at which the red and blue arcs cross? Here red marks the trajectory of point b, and blue that of point d.

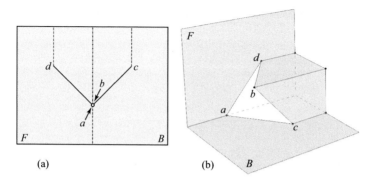

Figure 1.10 (a) Template for angled cuts. (b) The pop-up.

1.5 Cut Variations

We've examined symmetric and asymmetric cuts, but many other variations are possible. For example, angled but symmetric cuts (and still creases parallel to L), as in Fig. 1.10(a), lead to a very similar pop-up with parallel creases, as in Fig. 1.10(b).

Before we turn to popping up letters, the reader might enjoy exploring several other variants in the following exercises, perhaps verifying understanding with scissors and paper (cardstock is not essential for quick experiments).

Exercise 1.3 Understanding: Parallel Unequal

Describe the behavior of the pop-up whose template is shown in Fig. 1.11.

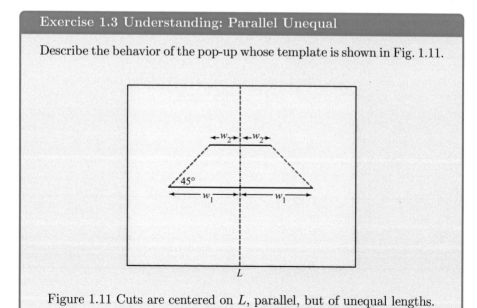

Figure 1.11 Cuts are centered on L, parallel, but of unequal lengths.

Exercise 1.4 Understanding: Three Parallel Cuts

Describe the behavior of the pop-up whose template is shown in Fig. 1.12.

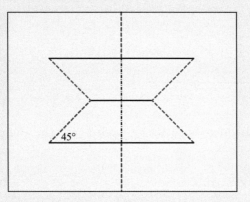

Figure 1.12 Three centered, parallel cuts.

Exercise 1.5 Understanding: Slanted Parallel

Describe the behavior of the pop-up whose template is shown in Fig. 1.13, with valley creases as shown.

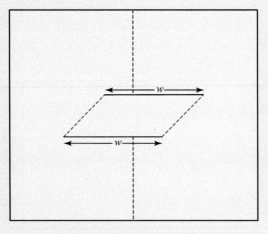

Figure 1.13 Parallel but slanted cuts.

1.6   Pop-Up Letters

It is now a short step from popping up a parallelogram box to popping up letters "carved" into the front of such a box. Fig. 1.14 illustrates the idea for the letter

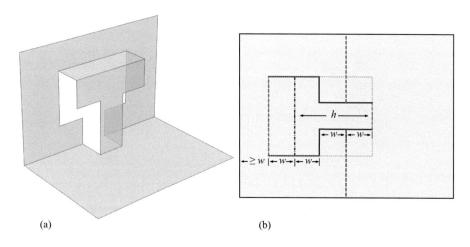

(a)   (b)

(c)

Figure 1.14 (a) Pop-up letter T and (b) template. (c) Several letters. Animation available for (a) (O'Rourke, 2021). [Pattern from Card3D software.]

T, which can be generalized for any sequence of letters. There is software all over the Web that creates a template for a given sequence of letters, and indeed we used software for Fig. 1.14(c).

Exercise 1.6 Practice: Tallest Letter

If a card has dimensions $W \times H$, what is the tallest letter one can pop-up, without having some of the letter protrude when the card is fully closed? (See also Exercise 1.1.)

1.7   Tents

So far we have explored constructions that pop-up cut portions of the front and back of the card. This constrains the structure so that the top (the w-side) of the popped-up parallelogram box (in Figs. 1.6 and 1.9) remains at all times parallel to the card back B and the h-side of the parallelogram remains parallel to the card front F. One can add structures not cut from the card, but instead

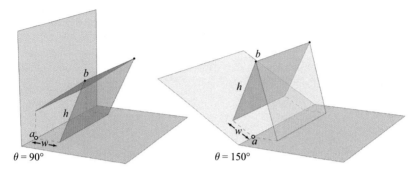

Figure 1.15 Tent, with $h = 3w$.

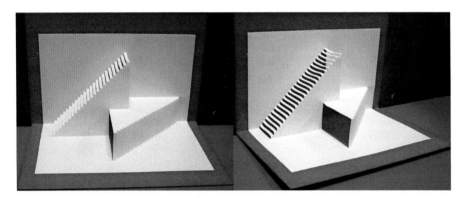

Figure 1.16 Two staircases. [Design and construction by Gail Parsloe. Based on construction in (Jackson, 1993, p. 63).]

separately creased cardstock pasted onto the card back and front. Fig. 1.15 shows a simple *tent* that does not flatten at $\theta = 180°$ as have our previous examples. When the card is fully opened at $\theta = 180°$, the tent has height

$$|ab| = \sqrt{h^2 - w^2}\,,$$

which is positive whenever $h > w$. We will examine similar pasted-in structures in the next chapter and especially in Chapters 5, 6, and 7.

1.8 Artistic Designs

Even with just the simple parallel fold, artistic designs can be made. Fig. 1.16 shows two intersecting staircases, one vertical and one horizontal. Each is constructed from parallel cuts staggered by the stair height. The extra layer of cardstock as a backdrop highlights the design.

Fig. 1.17 shows a beautiful design, again constructed solely of parallel cuts and a few "windows." One can view the central circular region as akin to the platform into which the T is cut in Fig. 1.14. We will see an analogous artistic example in Chapter 6, Fig. 6.1.

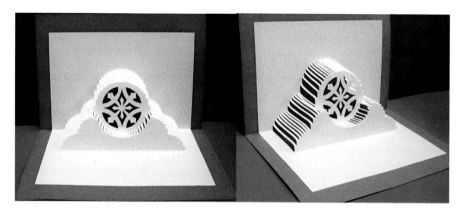

Figure 1.17 Logo for Evermore Origamic Architecture. [Construction by Gail Parsloe. Pattern from `origamicarchitecture.com`.]

Exercise 1.7 Challenge: Half-Cylinder

Design a pop-up that approximates half of a cylinder when opened to $\theta = 90°$, as in Fig. 1.18.

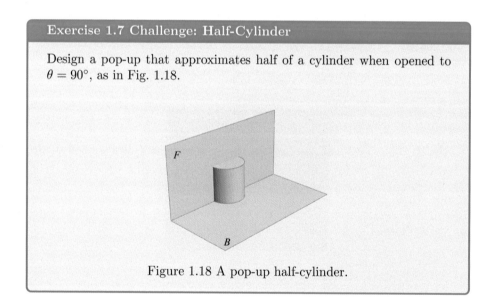

Figure 1.18 A pop-up half-cylinder.

1.9 Creases as Centerlines

A theme throughout this book will be identifying substructures in a pop-up design that have a simple relationship to the card angle θ during the opening motion. The simplest relationship is identity, and that is what is present in the pop-up parallelogram structure. As mentioned earlier, the top face of the parallelogram "box" always remains parallel to the card back B, and the lateral face of the parallelogram remains parallel to the front face F. So the valley creases where the parallelogram meets B and F (at points c and d in Figs. 1.5 and 1.9) are "copies" of the card centerline L, exhibiting the same dihedral angle θ.

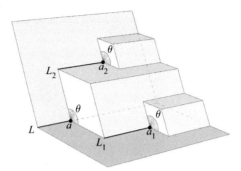

Figure 1.19 The card angle θ at a repeats at a_1 and a_2. Animation available (O'Rourke, 2021). [Design from (Jackson, 1993, p. 36).]

Figure 1.20 Fractal Christmas tree. [Design and construction by David Richeson; used by permission of author.]

This permits building another pop-up structure on those creases, viewing each crease as a displaced centerline. Fig. 1.19 shows one possibility. Here θ occurs at a along L, as well as at points a_1 and a_2 along L_1 and L_2, respectively. The pattern can be repeated, leading to the impressive fractal-like recursive pop-up in Fig. 1.20.

Notes

Impressive commercial software is available for designing letter (and more complex) pop-ups, e.g., POP-UP CARD DESIGNER PRO, tamasoft.co.jp/craft/popupcard-pro_en/.

For the Christmas tree design (Fig. 1.20), see Dave Richeson's blog, artfulmaths.com/blog/folding-christmas-fractals.

2

V-Folds and Rotary Motion

In this chapter we move beyond parallel folds and study the simple but powerful *V-fold*. The mathematics is a significant notch more complex, but opens new possibilities. A theme in pop-up design is to convert the card-opening motion—increasing θ—to drive other, rather different motions. So, the parallelogram pop-ups in Chapter 1 track circles centered on the card centerline L, and from an overhead view are "horizontal," parallel to the x-axis. The V-fold converts θ directly into rotation in the medial plane (Section 2.1.1). A greater change is that a V-fold opening motion can drive a "vertical" motion, overhead-parallel to the y-axis (Section 2.2). And an even more consequential transformation is converting θ into "flat" rotation in the xy-plane (Section 2.3).

We first explore the rich geometry of pure V-fold motion before turning to variations.

2.1 Geometry of V-Folds

2.1.1 Description of Motion

The basic V-fold, shown in Fig. 2.1, starts with the same symmetric cut as a parallel fold (Fig. 1.4), but diagonally creases to an apex point a on the card centerline L, making an angle α there. Let b and c be the cut endpoints on the back B and the front F, respectively, and p the midpoint of the cut. As in Chapter 1, we think of the card back B as fixed on a table top and the front F rotating from initially closed at $\theta = 0°$ to fully opened at $\theta = 180°$. Point c is fixed on and so moves with F. Point p moves with θ, and with it the two triangles $\triangle abp$ and $\triangle acp$. We will think of p's position in space as a function of θ, $p(\theta)$. Two particularly important locations of p are $p(0°)$ and $p(180°)$, p's position at the start and at the end of opening. We will label these points as p_0 and p_π, respectively. Although throughout this book we are measuring angles in degrees rather than radians, the notation $p_{180°}$ is a bit too cumbersome, so we use p_π instead, knowing that π radians is the same as $180°$.

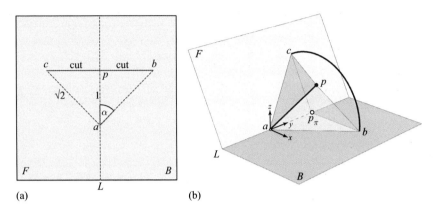

Figure 2.1 (a) V-fold template, $\alpha = 45°$. (b) V-fold at $\theta = 120°$. Point c follows the circle arc centered on p_π.

We start with four observations concerning the motions of the V-fold. First, it is a *one degree-of-freedom* motion, as in fact are all pop-up card motions. ("One degree-of-freedom" is abbreviated 1-dof.) This means that there is one parameter, in our case the centerline card angle θ, which determines all motions of parts of the pop-up construction. All parts are functions of θ; that angle drives all movements. It is not that other fixed parameters do not play a role. For example, the angle α affects the motion. But α is fixed in the construction.

Second, as mentioned, point c moves with the motion of F. So c follows an arc of a circle centered on p_π of radius $|cp_\pi|$, with the circle lying in a plane orthogonal to the centerline L. (Recall our notational conventions for segments and their lengths: Box 1.3.) This is illustrated in Fig. 2.1(b): c simply follows the opening of F, tracking θ on the illustrated circle.

From now on we focus on the motion of p, as that point determines the structure of the V-fold at any given stage of opening. The third observation is that p moves entirely in what we will call the *medial plane M*. This is a plane through the centerline L and rotated $\theta/2$ from the card back B. See Fig. 2.2. That p remains in the medial plane throughout the motion relies on the cut being symmetric about L, so that $|bp| = |cp|$.

Fourth, p follows a circular arc in the moving medial plane, centered on the apex a, with radius $|ap|$. See Fig. 2.3. This follows because the mountain crease segment ap has a fixed length, and a is fixed on the centerline L. So, given that p is in M, it must lie on that circular arc.

So far, we have made qualitative observations, but it will be informative to quantify the relations, using explicit coordinates and equations. To that end, let us fix the coordinate origin at a, with the xy-plane the fixed card back B and z the height above B; see again Fig. 2.1(b). Let us specialize the V-fold to $\alpha = 45°$, with the distance $|ap| = 1$, as illustrated in Fig. 2.1(a). With these conventions, we have fixed points and quantities, "fixed" in that they are independent of θ:

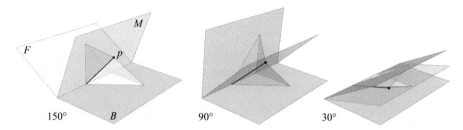

Figure 2.2 Point p lies in the medial plane M throughout its motion. Animation available (O'Rourke, 2021).

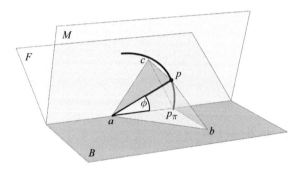

Figure 2.3 V-fold at $\theta = 150°$. The arc (blue) followed by p is in the medial plane M, where $\phi = \angle pap_\pi$.

$$a = (0,0,0)\,,\ b = (1,1,0)\,,\ p_\pi = (0,1,0)\,,$$
$$|ap| = |bp_\pi| = |ap_\pi| = 1,\ \ |ab| = |ac| = \sqrt{2}\,.$$

The points c and p have coordinates that depend on θ, and we now analyze each.

Position of c. As mentioned previously, c follows a circular arc and so, as we discussed in Chapter 1 (p. 7, Eq. 1.3), can be represented as

$$c = c(\theta) = (\cos\theta, 1, \sin\theta)\,,$$

which is the *parametric equation* for a circle in the 2D xz-plane. (See Boxes 2.1 and 2.2.) The expression $(\cos\theta, 1, \sin\theta)$ then represents a circle at $y = 1$, in a plane orthogonal to the card centerline L, and centered on p_π on L. So for $\theta = 0$, $c = (1,1,0) = b$: b and c are co-located when the card is closed. And when the card is fully opened at $\theta = 180°$, $c = (-1,1,0)$.

Box 2.1 Implicit Equations

There are three different ways of representing a curve in the xy-plane: direct, implicit, and parametric. The *direct* functional representation

expresses the y-variable as a function of x, $y = f(x)$. So, $y = x^2$ represents a parabola. But it is not always possible to isolate the y-variable
from its dependence on x. For example, a circle needs to be partitioned
into two functions: $y = +\sqrt{4 - x^2}$ and $y = -\sqrt{4 - x^2}$ for a circle of
radius 2. The reason is that $y = f(x)$ assumes that the curve intersects
a vertical line at x just once, but a circle intersects some vertical lines
twice; see Fig. 2.4.

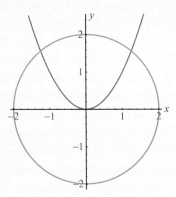

Figure 2.4 Parabola $y = x^2$ (blue) and circle of radius 2: $y = \pm\sqrt{2^2 - x^2}$
(orange and green).

An *implicit* representation provides an equation in x and y with
the understanding that precisely all pairs (x, y) that satisfy the equation
are points on the curve. The implicit equation for the parabola is still
$y = x^2$, but for the circle the implicit equation is $x^2 + y^2 = 4$.

A third representation is useful in many circumstances: *parametric
equations*. See Box 2.2.

Box 2.2 Parametric Equations

We used parametric equations in Chapter 1 (p. 7) in passing. Here we
explain more fully.

The idea is that, because a curve is one-dimensional, we can imagine
it being controlled by a single parameter determining x and y. For the
parabola, $x(t) = t^2$ and $y(t) = t$ map out points on the curve as a
function of the parameter t as it ranges from $-\infty$ to $+\infty$. For the circle
of radius 2, the parametric equations are

$$x(t) = 2\cos t \, ,$$
$$y(t) = 2\sin t \, ,$$

with the parameter t now ranging from 0 to 2π. This range is formally notated as $t \in [0, 2\pi]$, where $\in$ is the mathematical symbol for "is a member of" and $[0, 2\pi]$ represents the real numbers in the interval from 0 to 2π. Parametric equations are especially useful in computer graphics. In this book, we will freely mix curve representations, using whichever is more convenient.

Exercise 2.1 Practice: Parametric Ellipse

What are parametric equations for an origin-centered ellipse of major axis a (along the x-axis) and minor axis b (along the y-axis)?

Position of p. The motion of p is more difficult to describe quantitatively. One natural hope is that p's position along the circular arc in the medial plane M is some simple function of θ. Let ϕ be the angle between ap and ap_π, an angle lying in M; see Fig. 2.3. Then ϕ determines p, and so ϕ is some function of θ, with $\phi = 90°$ when $\theta = 0$ and $\phi = 0$ when $\theta = 180°$. However, the function is not simple. Under our assumptions ($\alpha = 45°$ and segment lengths detailed previously), with some effort one can calculate that

$$\phi = \cos^{-1}\left(\frac{1 - \cos\theta}{3 + \cos\theta}\right) , \tag{2.1}$$

an equation I find beautiful but hardly "simple." (See Box 2.3 for the meaning of $\cos^{-1}$.) So, when $\theta = 90°$, the $\cos\theta$ terms are 0, and we have

$$\phi = \cos^{-1}(1/3) \approx 70° .$$

In Fig. 2.3, $\theta = 150°$ leads to $\phi \approx 29°$. (The symbol $\approx$ means "approximately equal to.")

Box 2.3 Inverse Trigonometric Functions

Every trigonometric function has an inverse, which maps a length to the angle that yields that length. Because $\sin 30° = \frac{1}{2}$, the inverse sine of $\frac{1}{2}$ is $30°$. But notice that also $\sin 150° = \frac{1}{2}$, so sometimes one must choose which is the relevant inverse angle. Two notations are common, for example, for the inverse cosine: $\arccos\phi$ and $\cos^{-1}\phi$. We will use the latter.

2.1.2 Sphere Intersections

We now return to qualitative descriptions of the dynamic geometry of the V-fold motion, concentrating on the movement of p as θ varies. The physical constraints that follow from the V-fold construction require that p remains a distance 1 from the apex a, and also distance 1 from both b and c. The cardstock doesn't tear or rip, so the constraints that hold in the template—the fully opened state $\theta = 180°$ (Fig. 2.1(a))—hold throughout the motion. One way to represent these constraints on p is

$$|pa| = |pb| = |pc| = 1 \ .$$

If we express p explicitly as $p = (x, y, z)$, then these three distance constraints become three equations with three unknowns x, y, z. These equations can be solved; indeed they are solved every second around the world as part of GPS calculations; see Box 2.4. Before exploring these equations algebraically, let us turn first to the underlying geometry.

Box 2.4 GPS Trilateration

The Global Positioning System that pinpoints your phone's location works by intersecting three spheres whose radii are the distances to three satellites. These distances are computed by measuring the time for a speed-of-light signal to reach each satellite. As with the pop-up calculation, there are in general two solutions to the three equations. But generally, one is on the surface of the earth and the other is in the sky, so it is easy to dismiss the latter and use the former.

In actual practice, GPS uses four satellites, because beyond the three coordinates for position, the calculation needs to determine time as well, to convert the signals' time differences to spatial distances.

Exercise 2.2 Understanding: Intersection of Two Circles

Calculate the coordinates of the two points of intersection of two unit-radius circles in the xy-plane, one centered on $(0, 0)$ and the other centered on $(1, 0)$.

Each of the three distance equations represents a constraint of p lying on the surface of a sphere. Let S_a, S_b, and S_c be spheres centered on a, b, and c, respectively, each with radius 1. The intersection of S_b and S_c, which is notated as $S_b \cap S_c$, is a circle centered on the line segment connecting their centers b and c and lying in the medial plane M. See Fig. 2.5. The third sphere S_a crosses this circle at two points: p_π and p. Note that p_π on the centerline is a solution independent of θ, for p_π is a unit distance from a, b, and c. It is the other solution that yields the position of p as a function of θ.

Exercise 2.3 Understanding: Three Spheres, One Point

Describe conditions under which three unit-radius spheres, S_a, S_b, and S_c centered on a, b, and c, respectively, have just one point of intersection.

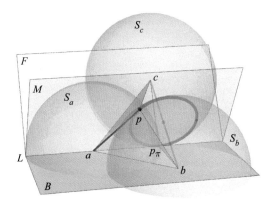

Figure 2.5 The three unit spheres intersect in two points: $S_a \cap S_b \cap S_c = \{p, p_\pi\}$.

The corresponding algebraic calculation of $p(\theta)$'s movements are detailed in Box 2.5.

Box 2.5 Solving Three Equations in Three Unknowns

Here we sketch how to solve explicitly the three equations in three unknowns representing the intersection of the three spheres:

$$|pa| = |pb| = |pc| = 1 .$$

The three unknowns are the coordinates of $p = (x, y, z)$. The distance between two points in 3D can be obtained by subtracting the two point coordinates to yield a vector between them and then expressing the length of that vector as the square root of the sum of the coordinate lengths squared, as explained in Box 1.3.

So the first equation $|pa| = 1$ is $|p - a| = 1$, with $a = (0, 0, 0)$:

$$\sqrt{(x - 0)^2 + (y - 0)^2 + (z - 0)^2} = 1 .$$

It is best to avoid the square root by squaring both sides:

$$x^2 + y^2 + z^2 = 1 . \tag{2.2}$$

The equation $|pb| = 1$, with $b = (1, 1, 0)$, is

$$(x - 1)^2 + (y - 1)^2 + z^2 = 1 . \tag{2.3}$$

The equation $|pc| = 1$, with $c = (\cos\theta, 1, \sin\theta)$, is

$$(x - \cos\theta)^2 + (y - 1)^2 + (z - \sin\theta)^2 = 1 . \tag{2.4}$$

Now one can use Eq. 2.2 to solve for z, then substitute that in Eqs. 2.3 and 2.4 to obtain two equations in x, y. Then solve Eq. 2.3 for y and

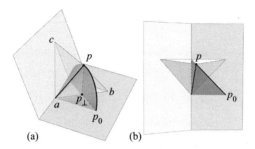

Figure 2.6 (a) The ap cone, $\theta = 120°$. (b) Overhead view, $\theta = 150°$.

substitute into Eq. 2.4 to finally obtain an equation in one variable x. This is not a simple calculation, but it results in a quadratic (degree-2) equation. We already noted that $(x, y, z) = (0, 1, 0)$ is a solution. The other solution is

$$x = 2 - \frac{4}{3 + \cos\theta},$$
$$y = -1 + \frac{4}{3 + \cos\theta},$$
$$z = \frac{2\sin\theta}{3 + \cos\theta}.$$

2.1.3 Cone Rim

We now turn to a rather different geometric view of the card-opening motion, which will play a role in Section 2.3 in generating flat rotary motion. Think of the motion of the segment ap as the card opens, with the card back B fixed as usual, but ignoring the medial plane. This segment sweeps out a cone, whose apex a is fixed to the card back B. See Fig. 2.6(a). The rim of the cone is the path of p in space, which is a circle arc centered on the point $p_\perp$ on the segment ab, where we are using the symbol $\perp$ to indicate the projection of p down the triangle $\triangle abp$ perpendicular to ab; so $pp_\perp$ is the altitude of that triangle. The radius of the circle arc is the length of that altitude, $|pp_\perp| = \sqrt{2}/2$ (for $\alpha = 45°$). The near-overhead view in Fig. 2.6(b) shows that the cone rim projects to the straight segment connecting p_0 to p_π. We will return to this viewpoint in Section 2.3.

2.1.4 Dihedral Angle $\delta = \theta$

Although we have seen some messy trigonometric equations describing the V-fold motion, there is one pleasing regularity in this motion: the dihedral angle δ along the mountain crease ap is exactly equal to the card angle θ. This dihedral

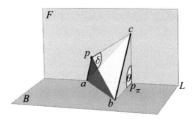

Figure 2.7 View into open V-fold: $\delta = \theta$.

angle δ is the angle in space between the two triangles hinged along ap: $\triangle abp$ and $\triangle acp$. Recall from Box 1.1 that the angle between two triangles is measured in a plane orthogonal to their shared edge, here ap. We now prove that $\delta = \theta$, in six steps, referring to Fig. 2.7.

1. The triangle $\triangle bcp_\pi$ is orthogonal to the centerline L. This is perhaps most evident in the V-fold template, Fig. 2.1(a).

2. Therefore the angle of that triangle at p_π is θ, which is of course the dihedral angle between B and F.

3. Both $\triangle abp$ and $\triangle acp$ are right triangles, with the $90°$ angle incident to p: they are right triangles in the template Fig. 2.1(a), and they retain their shape throughout the motion.

4. Therefore the angle $\angle bpc$ at p is the dihedral angle δ we seek.

5. The triangles $\triangle bcp_\pi$ and $\triangle bcp$ are congruent, because (a) they share edge bc, and (b) they are both isosceles with side lengths 1:

$$|bp_\pi| = |bp| = 1 \;\; , \;\; |cp_\pi| = |cp| = 1 \; .$$

6. Therefore the angles at p_π and at p are identical: $\delta = \theta$.

One can check that this reasoning does not rely on the particular choice of $\alpha = 45°$. However, it does rely on the V-fold cut being symmetric with respect to the centerline, i.e., we must have $|bp_\pi| = |cp_\pi|$ for the result to hold.

So we have proved this theorem:

Theorem 2.1 V-fold $\delta = \theta$

The dihedral angle δ at the mountain crease of a V-fold is equal to the card angle θ (independent of the V-fold angle α).

Exercise 2.4 Understanding: $\delta = \theta$ when $\alpha \neq 45°$

Argue that Theorem 2.1 holds for values of α different from $45°$.

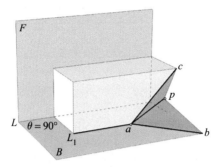

Figure 2.8 V-fold based on a valley crease L_1 parallel to the centerline L.

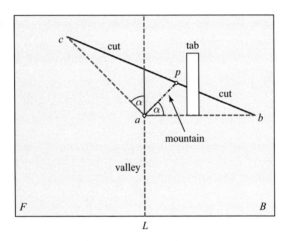

Figure 2.9 Template for y-motion. The pasted-in tab could be the start of a construction that moves vertically.

Before moving to the next section, we remark that, just as in Fig. 1.19, a V-fold can be based on a displaced valley crease L_1 parallel to the centerline L, as illustrated in Fig. 2.8. The construction in Section 2.3 will use a V-fold on the underside of a mountain crease, which is of course a valley crease.

2.2   Vertical y-Motion

Suppose that you want to design a pop-up card that includes a rocket lifting off straight up vertically as the card angle θ increases. The challenge is to convert the largely horizontal card-opening motion to vertical motion. Here interpret "horizontal" and "vertical" as in an overhead view, where z-motion is not relevant. This vertical y-motion can be achieved by a V-fold variation, whose template is shown in Fig. 2.9.

As in a V-fold, there is a single cut and then valley-mountain-valley folds to a point a on the centerline L. But, in contrast to Fig. 2.1(a), the cut is slanted and asymmetric. The two angles marked α must be equal for the abc triangle to fold flat at $\theta = 0$, and for the vertical motion we seek, $\alpha = 45°$. Thus segment ab is perpendicular to L at a, and it is $\triangle abp$ that rotates about ab and provides the vertical motion. Any structure pasted onto $\triangle abp$, such as the tab of cardstock shown, also rotates around ab and so remains horizontally fixed as its y- and z-coordinates change.

Consider the point p, the tip of the mountain fold along segment ap. As $\triangle abp$ rotates about ab, p follows the arc of a semicircle in the yz-plane, as illustrated in blue in Fig. 2.10. Unlike the median arc we saw in Fig. 2.3, along which p only traversed a total angle of about $70°$, here p follows a full $180°$ semicircle. Let $p_\perp$ be the center of the circle arc on ab and $r = |pp_\perp|$ its radius. Then when the card is closed at $\theta = 0°$, p_0 is on the card back B, at a distance r below the segment ab. And when the card is fully open at $\theta = 180°$, p_π is r above ab. If we again use ϕ to represent the angle of p around the circle arc, then ϕ travels from $180°$ to $0°$ as θ opens from $0°$ to $180°$. Although this suggests that perhaps $\phi = 180° - \theta$, in fact one can see in Fig. 2.10 that ϕ and θ do not agree at $\theta = 90$, when $\phi = 135°$. One can view the motion of p on its circular arc as being driven by the position of point c on the card front F. The cut segment cp is fixed in length, and so p must lie on a sphere S_c centered on c with radius $|cp|$. See Fig. 2.11.

Exercise 2.5 Practice: Sphere $\cap$ Circle

A sphere can intersect a circle in zero, one, two, or an infinite number of points. Which of these situations holds in Fig. 2.11?

Exercise 2.6 Challenge: Full p-arc

In order for p to traverse the full semicircle arc (blue) in Fig. 2.11, it must hold that the distance $|cp|$ is exactly long enough so that it can reach from c_0, the position of c when the card is closed, to p_0, the position of p at the beginning of the arc: $|cp| = |c_0p_0|$. Prove that this equality is realized by the design in Fig. 2.9.

Remarkably, the $\delta = \theta$ equality proved in Theorem 2.1 holds even for this angled V-fold—the dihedral angle δ along the mountain crease ap is exactly θ—a claim we will not stop to prove.

2.3 "Flat" Rotary Motion

Suppose now that you want to design a pop-up card that shows a cow jumping over the moon, or a card showing the arc of a basketball headed toward a hoop.

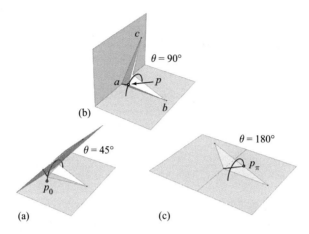

Figure 2.10 Snapshots of the y-motion. In overhead-view projection, p moves parallel to the y-axis along the card centerline. Animation available (O'Rourke, 2021).

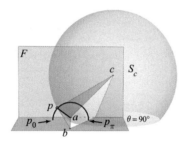

Figure 2.11 Point p lies on both the ap circular arc (blue) and the sphere S_c centered on c.

For these and many other similar scenarios, a mechanism is needed to convert the card opening by θ into a rotary motion flat in the plane of the card back B. This is a challenge, for, as we have seen, θ operates in a plane orthogonal to the centerline L, whereas the desired rotary motion is in the B-plane (or a plane closely parallel to B). These are fundamentally different rotations, and it is not immediately evident how to convert the card-opening motion to B-plane rotation. Nevertheless, many card designs accomplish this. However, I do not know if it is possible to effect an <u>exact</u> conversion, similar to our exact vertical motion in the previous section. Instead, in this section we will explain how to use a V-fold to perform an approximate conversion, good enough to support most designs.

Recall from Fig. 2.6(b) that, in a standard symmetrical V-fold, as p travels from p_0 (card closed) to p_π (card fully opened), the projection of its cone-rim riding motion onto B is the straight segment $p_0 p_\pi$. In Fig. 2.12 we map out the projection of the segment bp, and one can see that it rotates about b, in

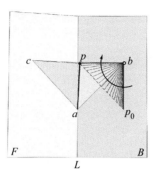

Figure 2.12 Rotation of bp centered about b in projection.

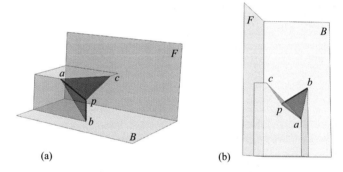

Figure 2.13 (a) Inverted V-fold. (b) The segment bp will again rotate about b in projection.

projection. Let us call this angle β, the angle with respect to the start position bp_0. In the figure, β rotates 90° during the full card-opening motion. This is the "engine" behind the conversion from θ to β. But there is a challenge in that this "clean" rotation is only clean in projection to 2D, whereas the 3D motion is more complex: the segment bp sweeps out a cone apexed at b, whose cone rim is identical to the rim of the cone apexed at a that we saw in Fig. 2.6(a). We now describe a way to meet this challenge.

First, we start with a displaced parallel fold, producing a type of shallow table. Second, we use a V-fold centered on the mountain crease of this table, indented inward, so that instead of ap being a mountain crease, as in Fig. 2.8, it becomes a valley crease. See Fig. 2.13(a). The labeling and geometric behavior of the V-fold is the same as we used in Section 2.1.

In this example, we have exaggerated the z-height of the table, so that the mechanism will be evident. Typically the table top is lower, closer to B. Looking overhead in Fig. 2.13(b), we will follow the sweep of the segment bp, which because of the inversion of the V-fold, is now close to B. In projection,

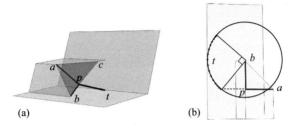

Figure 2.14 (a) A segment *pt* attached to *bp*. (b) Overhead view when $\theta = 0$. The initial position of *pt* is shown dashed. Animation available (O'Rourke, 2021).

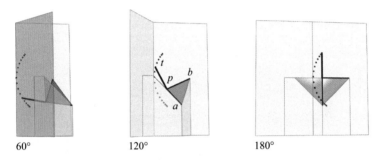

Figure 2.15 Rotary motion of the *pt* stick at several values of θ. Animation available (O'Rourke, 2021).

again this segment will rotate about *b* (as we saw in Fig. 2.12), but again, the motion is not in the plane of *B*, but rather in 3D.

Now here is the clever design trick to achieve the desired rotary motion in the plane of *B*. Imagine attaching a "stick" segment *pt* that (a) is at all times perpendicular to *bp* and (b) touches the plane *B* at its *t* endpoint. See Fig. 2.14(a). As Fig. 2.14(b) shows, the points at which *t* rests on *B* as the card is opened map out a near-perfect circular arc on *B*. It is not perfect because the *p* endpoint of the stick is riding a cone rim above *B*. Several snapshots are shown in Fig. 2.15. So we have now converted the 180° rotation of θ about the card centerline into a 90° rotation in the plane *B* of the card back. One can notice that the dots in Fig. 2.14(b) are not quite uniformly spaced, despite being generated at equal θ intervals. So the relationship between the two angles is not as simple as β being exactly half of θ. Nevertheless, it is a sufficient approximation for most needs.

In place of the stick *pt*, typically a design just extends the cardstock beyond *bp* (or attaches a cardstock tab to *bp*). An example template is shown in Fig. 2.16. The pink triangle $\triangle abp$ is joined to the blue disk. The resulting card, shown in Fig. 2.17, has a pleasing dynamic, which can be employed to drive specific artistic rotary designs.

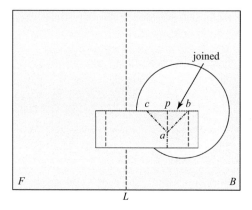

Figure 2.16 Rotary motion template.

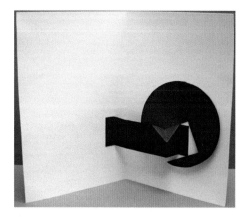

Figure 2.17 The card whose template is shown in Fig. 2.16.

Exercise 2.7 Practice: V-fold angle $\alpha \neq 45°$

We have used the V-fold angle $\alpha = 45°$ throughout (Fig. 2.1(a)), which led to a 90° rotation in Fig. 2.15. How much rotation will occur with V-fold angles larger or smaller than 45°?

Notes

What I call a V-fold is called an "angle fold" in (Carter and Diaz, 1999). Asymmetric angled V-folds (Section 2.2) are discussed in (Jackson, 1993, p. 32).

3

The Knight's Visor

3.1 The Knight's Visor Pop-Up

In the previous two chapters, the basic parallel fold and V-fold used cuts that are perpendicular to the card centerline. One might think not much more can be achieved by such highly constrained cuts, but there is a beautiful variation resulting from many parallel cuts that forms (at $\theta = 90°$) a shape that I call the *Knight's Visor*. See Fig. 3.1 for the pop-up and Fig. 3.2 to justify the moniker. Unlike the rotary motion explored in Chapter 2, here it is the final shape that is of primary interest, although its emergence upon opening is quite pleasing; see Fig. 3.3. We will use this example to derive equations describing the shape, leading to connections with some beautiful classical mathematics. The derivations get a bit intricate—no calculus, just algebra—but despite their intricacy, the pursuit is worthwhile. The reader uninterested in the details might skim the derivations.

The design is remarkably simple. Draw a circle C centered on the card's centerline, and cut equally spaced, parallel cuts terminating on the circle boundary. See Fig. 3.4. This is most easily accomplished by first folding the card in half, and cutting with scissors from the centerline to the semicircles on the front and back of the card. The creases across the diameter of C are mountain creases. Each strip ends on the circle, where it joins there in a valley crease. These valley creases are straight segments that approximate the circle tangent; the creases are <u>not</u> parallel to the card centerline.

It will be convenient to change the coordinate system from what we followed in the previous two chapters so that (a) the x-axis is along the card centerline, and (b) the origin is at the center of centerline, which coincides with the center of the circle. See Fig. 3.5.

As the card opens and closes, the motion is quite intricate. As we will see in more detail later, each half-strip (a "rib") attached to the back of the card rides on a cone (see ahead to Fig. 3.13) whose axis is along its angled crease, and the half-strip rib attached to the front face of the card rides on another cone, but this time attached to the front face. And all the rib cone axes make

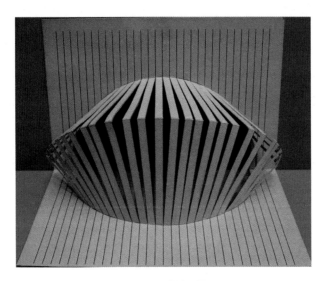

Figure 3.1 The Knight's Visor pop-up.

Figure 3.2 Knight's helmet with visor. [By permission of SwordsandArmor.com.]

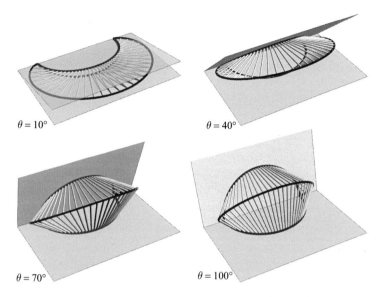

Figure 3.3 The Knight's Visor at several θ values. The rim is red, and the circle C is blue. Animation available (O'Rourke, 2021).

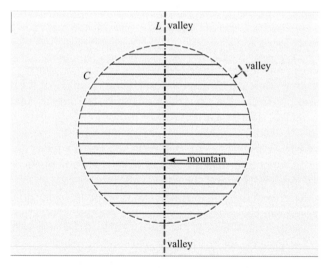

Figure 3.4 Template for the Knight's Visor. Each trapezoidal rib is valley-creased tangent to C.

different angles with the centerline. The combination produces a bowing that is small for the longer strips and gradually increases for the shorter strips. The result is a pleasant surprise as the card gradually closes from fully opened (card angle $\theta = 180°$), a shape variation difficult to predict from its mundane start

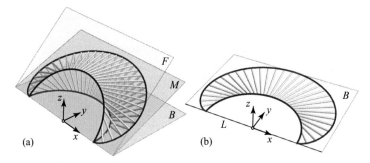

Figure 3.5 Card notation, with the x-axis along the centerline L. (a) At $\theta = 45°$, the rim is in the medial plane M. (b) At $\theta = 0°$, the rim is on the card back B.

in the template seen in Fig. 3.4. With thin ribs carefully cut and creased in quality cardstock, the visor shape is almost a work of art with the angle fixed to $\theta = 90°$.

Exercise 3.1 Understanding: Rim at $\theta = 90°$

When $\theta = 90°$, which point p of the rim is farthest from the card centerline L, and what is p's distance to L? Consult Fig. 3.3.

3.2   Flat Visor Curve

Call the midcurve of the knight's visor its *rim*. The symmetry of the construction about the centerline L ensures that the rim lies in the medial plane M at all times; see Fig. 3.5(a). The rim originates from the portion of the centerline of the card within the disk D that has been cut with parallel slits into *ribs*, each rib R a thin trapezoid between the card centerline and the boundary circle C of the disk. We know the shape of the rim when the card is fully opened ($\theta = 180°$): it is a diameter of C, right along the card centerline.

Our first goal is to derive the shape of the rim when the card is fully closed ($\theta = 0°$), which we call the *flat visor curve*, "flat" because the visor is flattened with the card closure. This curve is not visible (unless the card is transparent), but it is here that we connect to classical mathematics. In Section 3.4, we will turn to calculating the 3D shape of the rim as θ opens.

We first derive parametric equations for the flat visor curve, $p(s) = (x(s), y(s))$ in a parameter s corresponding to the x-coordinate of a rib. This is achieved in Eq. 3.6. Then in Section 3.2.2 we derive an implicit equation, achieved in Eq. 3.8, whose form connects to the classical mathematics representation of the nephroid curve (Section 3.3).

3.2.1   Parametric Equations

Because the valley crease where a rib meets C follows the contour of C and is a straight-segment crease, that crease is tangent to C. When the card is

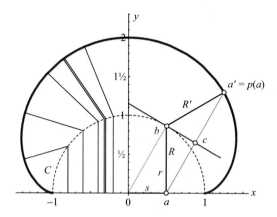

Figure 3.6 The yellow strips show two positions of a rib: card fully opened
($\theta = 180°$) and fully closed ($\theta = 0°$). The rib $R = ab$ reflects across the tangent
at b to $R' = ba'$.

fully opened ($\theta = 180°$), each rib is perpendicular to the diameter of C along
the card centerline. When the card is fully closed, each rib has rotated to lie
completely external to the disk D whose boundary is C. See the yellow strips in
Fig. 3.6.

Now we establish notation for various points to allow us to derive parametric equations for the rim on the card back, when $\theta = 0°$. (Parametric
equations are explained in Box 2.2.) Let C have unit radius and be centered
on the origin; see again Fig. 3.5. We will reduce each rib R to a segment,
$R = ab$, with a on the centerline, $a = (s, 0)$. We seek the point a' on the
flat visor curve corresponding to a, as a function of the parameter s, where
s ranges from -1 to 1, i.e., $s \in [-1, 1]$. The plan is to reflect a over the
tangent at b to a'. The one parameter s will eventually describe the x- and
y- coordinates of a'.

We only need one half of C, which can be described by the equation $y = \sqrt{1 - x^2}$, and so $b = (s, \sqrt{1 - s^2})$. Let r be the length of the rib R,

$$r = |R| = \sqrt{1 - s^2} \; ; \tag{3.1}$$

so the coordinates of b are $b = (s, r)$. Next we find the tangent to C at b, and
use that to determine the *reflected rib* R', reflected because the crease of the
rib R is tangent to C where it attaches at b. Refer throughout to Fig. 3.6.

Box 3.1 Perpendicular Slopes

That perpendicular lines have slopes that are the negative of the reciprocal of one another can be seen by rotating a slope triangle $90°$, as
illustrated in Fig. 3.7.

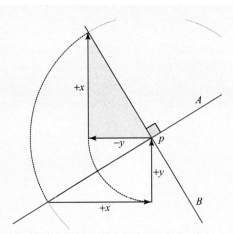

Figure 3.7 Rotating the blue slope triangle $+x$ and $+y$ about p by $90°$ clockwise results in the purple slope triangle $-y$ and $+x$.

We can find the slope of the tangent line to C by using the fact that the tangent is perpendicular to the radial segment from the origin to b. Because the slope of that radial segment is $\frac{r}{s}$, the slope of the tangent at b is $-\frac{s}{r}$. Here we use the fact that perpendicular lines have slopes that are the negative of the reciprocal of one another; see Box 3.1. From this we can derive an equation for the tangent line:

$$y = (s - x)\frac{s}{r} + r .$$ (3.2)

Next we compute the reflected rib R', by computing the reflection a' of a across this tangent line.

> ### Exercise 3.2 Practice: Reflection
>
> Let a line ℓ pass through the points $(1, 0)$ and $(0, 1)$. What is the reflection in ℓ of three points?
>
> - $(0, 0)$.
> - $(\frac{1}{2}, 0)$.
> - $(\frac{1}{2}, \frac{3}{2})$.

The segment aa' is perpendicular to the tangent and so has slope $\frac{r}{s}$, and therefore the line containing that segment has equation

$$y = (x - s)\frac{r}{s} .$$ (3.3)

Intersecting this line with the tangent line by solving Eqs. 3.2 and 3.3 simultaneously yields coordinates for the midpoint $c = (c_x, c_y)$ of the segment aa':

$$c_x = \frac{s(s^2 + 2r^2)}{s^2 + r^2} ,$$

$$c_y = \frac{r^3}{s^2 + r^2} .$$

$$(3.4)$$

Using $s^2 + r^2 = 1$ (recall that C has unit radius) and the definition of r (Eq. 3.1) reduces c to

$$c_x = s(2 - s^2) ,$$

$$c_y = (1 - s^2)^{3/2} .$$

$$(3.5)$$

Now we can compute $a' = (a'_x, a'_y)$ knowing that c is the midpoint of aa': $c = \frac{1}{2}(a + a')$ and $a' = 2(c - a) + a$, so

$$a'_x = x(s) = s(3 - 2s^2) ,$$

$$a'_y = y(s) = 2(1 - s^2)^{3/2} .$$

$$(3.6)$$

We have now arrived at parametric equations for the flat visor curve. Each $s \in [-1, 1]$ leads to a point $p(s) = (x(s), y(s))$ on the curve. Note that for $s = \pm 1$, $p(s) = (\pm 1, 0)$, and the curve is tangent to the x-axis there. For $s = 0$, $p(s) = (0, 2)$, corresponding to the centermost rib of length $r = 1$ reflecting across the horizontal tangent at the top of the circle C.

> **Exercise 3.3 Practice: Verify r^3**
>
> Verify in Fig. 3.6 by rough measurement Eq. 3.4: The y-coordinate of point c is r cubed: $c_y = r^3$. Then verify this by explicit calculation using $a = (\frac{1}{2}, 0)$, as illustrated in the figure.

3.2.2 Implicit Equation

We make it our next goal to obtain an implicit form of the equation of the flat visor curve. Recall from Box 2.1 that an implicit equation in x and y represents all the (x, y) pairs that satisfy the equation. So the goal is to obtain equations whose only variables are x and y, by eliminating the parameter s from the parametric equations. First, rewrite those equations (Eq. 3.6) in terms of r via Eq. 3.1:

$$x = 2sr + s ,$$

$$y = 2r^3 .$$

Solving the second expression for r yields $r = (y/2)^{1/3}$. Substituting that into the first gives

$$x = 2^{2/3} s y^{1/3} + s ,$$

and solving this for s yields

$$s = \frac{x}{1 + 2^{2/3} y^{1/3}} .$$

$$(3.7)$$

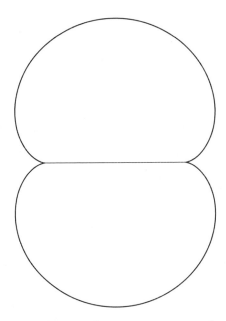

Figure 3.8 Reflecting the visor curve over the x-axis.

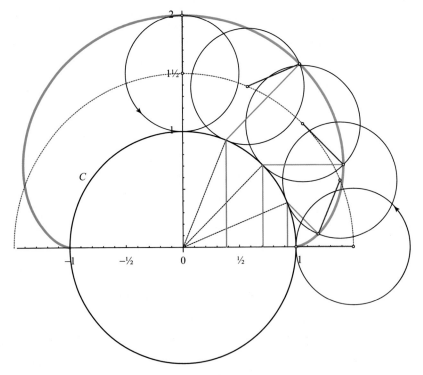

Figure 3.9 The circle of radius $\frac{1}{2}$ rolls without slippage on C.

Recall that $s^2 + r^2 = 1$. Substituting Eq. 3.7 for s and Eq. 3.1 for r into this circle equation yields

$$\frac{x^2}{\left(2^{1/3}y^{2/3}+1\right)^2} + \frac{y^{2/3}}{2^{2/3}} = 1 \, ,$$

$$2^{2/3}x^2 + \left(2^{1/3}y^{2/3}+1\right)^2 y^{2/3} = 2^{2/3}\left(2^{1/3}y^{2/3}+1\right)^2 \, ,$$

$$2^{2/3}\left(x^2+y^2\right) = 3y^{2/3} + 2^{2/3} \, ,$$

and finally

$$x^2 + y^2 = 3(y/2)^{2/3} + 1 \, ,$$

$$(x^2+y^2-1)^3 = \frac{27}{4}y^2 \, . \tag{3.8}$$

Note that now the only variables in these two equivalent equations are x and y.

3.3 The Nephroid

The last form of the implicit equation in Eq. 3.8, reveals it to be a *nephroid*, a planar curve studied since the seventeenth century. A nephroid is usually represented in the form

$$(x^2 + y^2 - 4t^2)^3 = 108t^4y^2 \, ,$$

with a parameter t. For $t = \frac{1}{2}$, $4t^2 = 1$ and $108t^4 = \frac{27}{4}$, and we reach exactly Eq. 3.8. The name of this curve derives from the Greek *nephros*, which means "kidney," for when the curve is drawn to both sides of the x-axis (or on both the card front and card back), it is indeed kidney-shaped: see Fig. 3.8.

The nephroid is a rich curve with several interpretations. For example, it is the trace of a point on one circle as it rolls on another; see Fig. 3.9.

Exercise 3.4 Understanding: Epicycloid

Referring to Fig. 3.9, suppose that the rolling circle has unit radius, the same as C. How many times does the curve traced by a point on the rolling circle touch the inner circle C ? In Fig. 3.9, the nephroid touches twice, at $(\pm 1, 0)$.

The interpretation most natural in our context is that our nephroid is the envelope of circles centered on C and tangent to the diameter of C on the x-axis. Fig. 3.10 shows that, for each of these circles, a pair of radii are the two ribs R and R': R perpendicular to the x-axis, and R' perpendicular to the visor curve, where the circle is tangent to the curve.

We mention one more interpretation of the nephroid. If the ribs R' in Fig. 3.10 are reflected inside C rather than reflected outside, then we can view the collection of ribs R as parallel light rays reflecting off an interiorly mirrored

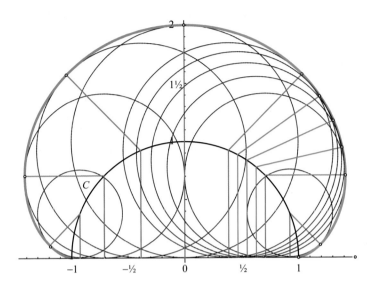

Figure 3.10 The flat visor curve is the envelope of circles centered on C and tangent to the x-axis. Several rib-radii are shown (in green).

circle C. The envelope of these reflected rays is a rotated nephroid of half the size, as illustrated in Fig. 3.11. We do not pause to derive this connection, but only marvel at it. The concentration of crossing reflected light rays forms a caustic (or catacaustic), which—remarkably—can be seen in a drinking glass or a tea cup under the right circumstances, as demonstrated in Fig. 3.12.

3.4   Visor Curve in 3D

Next we derive an equation for the visor curve in 3D. We seek parametric equations for points on the curve $p(\theta, s) = (x, y, z)$ as a function of both the card angle θ and the same parameter s we used for the flat visor curve, which identifies a particular rib. The geometry is a bit intricate, and the algebra even more so, but worth analyzing. We will see that $p(\theta, s)$ can be viewed geometrically as a point of intersection of two circles in space, one the intersection of two spheres and the other the intersection of a cone with a plane.

It is clear that the visor curve lies in the medial plane M (see Fig. 3.5(a)), which makes an angle $\frac{1}{2}\theta$ with the card back B. Because z/y is $\tan(\frac{1}{2}\theta)$, the z-coordinate of p is determined by its y-coordinate:

$$z(\theta, s) = \tan(\tfrac{1}{2}\theta)\, y(\theta, s) \ .$$

As usual, view the card back B as fixed in the xy-plane while the card front rotates. Now imagine fixing s and so fixing $a = (s, 0)$ and b. Because $p = p(\theta, s)$ is connected to point b by the rib R, p is a distance $r = |R|$ from b, and so it lies on a sphere S of that radius centered on b.

Let b_F be the analog of point b but on the card front instead of the card back. Then again p lies a distance r from b_F, on sphere S_F, which moves as θ

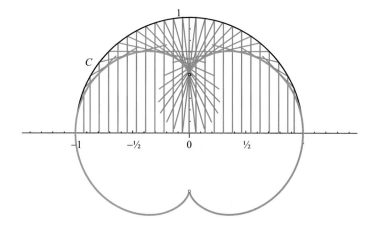

Figure 3.11 Caustic formed by parallel light rays reflecting inside C.

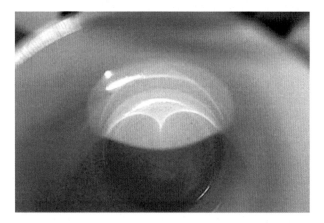

Figure 3.12 Caustic in a tea cup. [Wikimedia Commons, photo by Paul Venter.]

changes. The spheres S and S_F intersect in a circle C_M that lies in M. Refer to Fig. 3.13. We can represent the two-spheres constraint by requiring that the distance from p to each of b and b_F be r, the length of the rib:

$$|pb| = |pb_F| = r = \sqrt{1 - s^2}, \tag{3.9}$$

where $b_F = (s, r\cos\theta, r\sin\theta)$.

One more constraint will pin down $p(\theta, s)$. Returning to Fig. 3.6, note that the rib in 3D at all times forms the same angle with the tangent to C at b, because the crease at the base of the rib lies along that tangent. Thus as the rib rotates in 3D, attached all the while to b, it sweeps out a cone whose rim is a semicircle lying in vertical plane V, which is perpendicular to the xy-plane.

Recall that we earlier derived the equation for the line containing aa' in Eq. 3.3, which is the diameter of the semicircle traced by the rib tip.

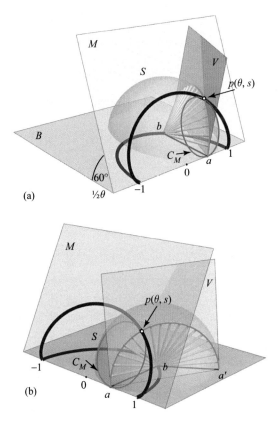

Figure 3.13 Two views of the Knight's Visor at $\theta = 120°$. Not shown: card front F, point b_F, and sphere S_F. Circle C_M (brown) is the intersection of S and S_F. Plane V cuts S in a semicircle (green) that passes through p.

So now we can mix two constraints to determine $p(\theta, s)$: (1) p lies on the two spheres given by Eq. 3.9, and so on their circle of intersection C_M in the medial plane M; and (2) p lies on the vertical plane V through the segment aa', which lies on a line whose equation is Eq. 3.3. The plane V cuts the cone in another circle. These two circles meet at two points, at a and at $p(\theta, s)$.

Solving Eqs. 3.9 and 3.3 simultaneously yields, after (considerable) simplification,

$$x(s, \theta) = -\frac{s\left((s^2 - 2)\cos(\theta) + 3s^2 - 4\right)}{s^2\cos(\theta) - s^2 + 2},$$

$$y(s, \theta) = \frac{4\left(1 - s^2\right)^{3/2}\cos^2\left(\tfrac{1}{2}\theta\right)}{s^2\cos(\theta) - s^2 + 2}, \tag{3.10}$$

$$z(s, \theta) = \tan\left(\tfrac{1}{2}\theta\right) y(s, \theta).$$

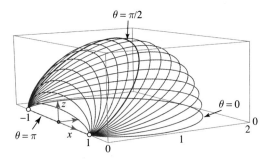

Figure 3.14 Visor rim curves over all θ.

These parametric equations were used to plot the curves in Fig. 3.3 and in the alternate view shown in Fig. 3.14. The $\theta = 90°$ curve corresponds to the rim in the photo in Fig. 3.1.

Although the equations in Eq. 3.10 look complicated, substitution of a fixed θ of course simplifies them. For example, with $\theta = 60°$, they reduce to

$$x(s, 60°) = s(5 - \tfrac{7}{2}s^2) \,,$$
$$y(s, 60°) = 3(1 - s^2)^{3/2},$$
$$z(s, 60°) = \sqrt{3}(1 - s^2)^{3/2} \,,$$

which has a form recognizably similar to the parametric equations (Eq. 3.6) we derived for the flat visor curve.

Exercise 3.5 Understanding: 3D Parametric Equations

Verify by substitution for θ in Eq. 3.10 the two extremes:

- $\theta = 0°$, where the curve is the flat visor curve (Eq. 3.6);
- $\theta = 180°$, where the curve reduces to the segment diameter of the circle C.

3.5   Parabolic Visor

We close this chapter with a beautiful variation on the Knight's visor pop-up due to Jean-Jacques Dupas. If, instead of slicing the card into parallel ribs anchored on a circle C, we slice to a parabola P—or rather facing parabolic arcs—then what was the nephroid curve when $\theta = 0°$ becomes an exact circle! A model, prepared with the help of a cutting machine, is shown in Fig. 3.15.

Figure 3.15 Parabolic visor pop-up. [Construction by Jean-Jacques Dupas; by permission of author.]

Box 3.2 Parabola

A parabola with focus at $(0, f)$ has equation $y = \frac{1}{4f}x^2$ and directrix $y = -f$. Each point p on the parabola is equidistant from the focus and the directrix. In Fig. 3.16, $f = 1$.

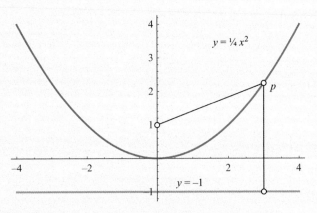

Figure 3.16 Parabola with focus at $(0, 1)$.

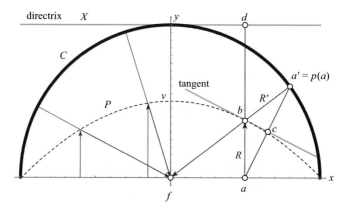

Figure 3.17 The segment ad and the segment fa' are of equal length. See Fig. 3.6.

Although it is more difficult to construct this pop-up, it has beautiful geometry behind it. In fact, the proof that the flat curve is a circle needs no algebra, just a few properties of parabolas. This might in some way compensate for the algebra-heavy proofs presented in the previous sections. In particular:

(1) Light rays parallel to a parabola P's axis reflect through the focus f. This is of course the property behind the parabolic reflector telescope. This reflection property has been known since 200 BC.

(2) The distance from a point p on the parabola to the focus f is the same as the distance from p to the directrix X. This is the defining property of a parabola. See Box 3.2.

Using these properties, we now detail a geometric proof that with the ribs defined by parabolic arcs, the flat curve is a circle. We refer throughout to Fig. 3.17, which parallels Fig. 3.6 used in Section 3.2.

Let P be the parabolic arc on the card back with its focus f on the card centerline, the x-axis. The directrix X is parallel to the x-axis and sits twice the distance $|fv|$ above v, where v is the "vertex" of P, in our setting the y-highest point of P. As before, let $R = ab$ be a rib. Rib R reflects over the tangent to P at point b, to $R' = ba'$, and it is $a' = p(a)$ that lies on the flat curve C. Our goal is to show that C is a circle centered on f.

Viewing R as a light ray entering from below at a and reflecting off P at b, we know from property (1) that the ray reflects through the focus f. Because the same tangent is responsible for R reflecting inward through f and R reflecting outward to R', we know that bf and ba' are collinear.

We now argue that the distance $\rho = |fa'|$ is the same regardless of the position of a, which will prove that C is a circle. For any a on the x-axis, the vertical distance $|ad|$ to the directrix X is the same, because X is parallel to the x-axis. Partitioning ad where it crosses P yields $|ad| = |ab| + |bd|$.

Next, look at fa', which is a segment because bf and ba' align, and partition fa' where it crosses P. The two partitions give us

$$|ad| = |ab| + |bd| = |R| + |bd| \,,$$
$$|fa'| = |fb| + |ba'| = |fb| + |R'| \,.$$

(3.11)

Now, by property (2) we know that the distance to the directrix and the distance to the focus are equal: $|bd| = |fb|$. And we know that $|R| = |R'|$ because R' is R reflected. So finally, substituting these equalities into the expressions in Eq. 3.11 shows that they are equal: $|ad| = |fa'|$. And so $\rho = |fa'|$ is a constant because $|ad|$ is fixed independent of a. In fact, the segment ad and the segment fa' are reflections of one another in the tangent at b. Therefore, the trace of all a' points is a circle C of radius ρ centered on f.

Exercise 3.6 Practice: Parabola Equation

What is the equation of the parabola P in Fig. 3.17, assuming that the focus is $f = (0,0)$, the parabola's vertex is $v = (1,0)$, and the directrix is $y = 2$?

Notes

Much of this chapter is based on (Jakus and O'Rourke, 2012). The original visor design is described in (Jackson, 1993, p. 62, Fig. III) as a "multi-slit variation." See (Wells, 1991, p. 158) for a derivation of the half-nephroid caustic in Fig. 3.11. Jean-Jacques Dupas invented the parabolic visor after reading (Jakus and O'Rourke, 2012). The geometric proof is due to François Lavallou.

4

Pop-Up Spinner

4.1 Pop-Up Spinner

The most stunning and elegant pop-up effect I have encountered is the *pop-up spinner* card, invented in Japan by an unknown student at Musashino Art University in 1988. Its nested diamond frames spin about a central axis as the card opens, with the inner frames spinning faster than the outer frames, creating a dazzling dynamic effect, as depicted in Fig. 4.1. When the card is fully opened flat, one sees a simple pattern of nested diamonds (squares tilted at 45°) cut directly into the card—unlike other impressive pop-up effects, this one does not rely on complex attachments, but is instead cut from just one piece of cardstock.

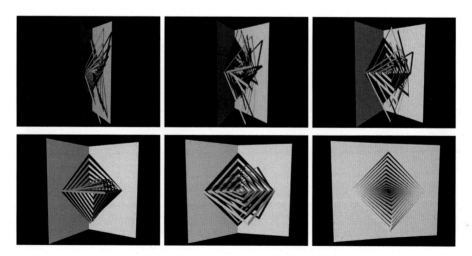

Figure 4.1 Pop-up spinner card opening (O'Rourke, 2011, Fig. 3.10). Animation available (O'Rourke, 2021). [Reprinted by permission of Cambridge University Press. Animation by Akira Nishihara.]

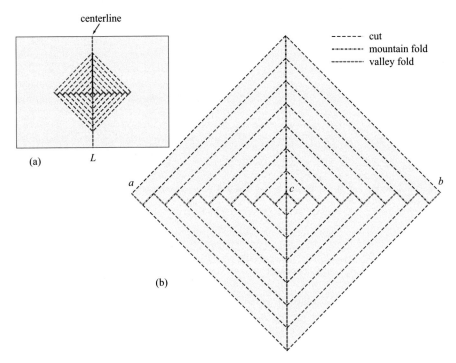

Figure 4.2 (a) Pop-up spinner card design. (b) Nested diamond details: cuts and folds (O'Rourke, 2011, Fig. 3.11). [Reprinted by permission of Cambridge University Press.]

How the card-opening motion is converted to spinner dynamics is by no means obvious. It turns out that the heart of the pop-up spinner, and the key to understanding its operation, is a theorem about linkages proved in an undergraduate thesis, the Unit 90°-Chain Theorem (Theorem 4.1). We will first explain the spinner construction in detail, and then prove the theorem.

Fig. 4.2 shows a template for the construction of a pop-up spinner. We encourage the reader to cut the pattern on cardstock and crease as indicated— no description can fully convey the beauty of the physical action. Box 4.1 walks through the construction.

Box 4.1 Cut and Fold the Pop-Up Spinner

Either draw the pattern with a ruler on cardstock, or download the template from this book's website (O'Rourke, 2021) and print on the heaviest weight paper your printer will accept. First crease the card down the card centerline, one long mountain fold. Don't worry that at this point some of the valley folds along that midline are folded backward. Now cut the diamond diagonal cut lines with scissors, stopping just at the central zigzag, as in Fig. 4.3. Fortunately, scissors suffice: no razor knife is need as in many intricate constructions. Reverse the folds along the card centerline that should be valley folds. Lay the design flat. Now, starting at point a in Fig. 4.2(b) and proceeding across to b, crease the zigzag segments with your fingers, alternating valley/mountain as indicated. Once every crease is folded the correct way, you can sharpen the creases with additional pressure. At this point you should be able to start twisting and compressing from the center point c outward, reversing the animation snapshots in Fig. 4.1. Then open: voilá!

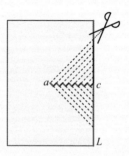

Figure 4.3 Cutting the pop-up spinner template from Fig. 4.2(a) (O'Rourke, 2011, Fig. 3.12). [Reprinted by permission of Cambridge University Press.]

Fig. 4.2 is too complicated for grasping the structure, so we will analyze portions of it separately. At a first glance, it is not evident how it all hangs together as one connected piece. Fig. 4.4(a) shows the cuts for just two of the nested square diamonds, revealing clearly that a horizontal strip remains free of cuts. Fig. 4.4(b) shows how a zigzag of mountain/valley creases runs through the uncut strip, which in Fig. 4.2 is seen to connect the left and right tips of the outermost diamond. Call these tip points a and b.

Now we set aside the pop-up spinner for a detour into linkages, returning to the spinner in Section 4.3.

4.2 Linkages and Fixed-Angle Polygonal Chains

The Unit 90°-Chain Theorem that drives the functioning of the pop-up spinner concerns configurations of particular linkages, which we now explain.

A *linkage* is a 3D mechanism built from rigid links sharing *joints*. Each link is a line segment whose length is fixed throughout the motion of the linkage, and no two links can collide or cross. Often such links are called *bars* to emphasize their rigidity. Although linkages are quite general and sometimes quite complicated mechanisms, here we will only need what are called *polygonal chains*: a sequence of links joined end to end, forming a chain C. So each joint connects exactly two links. Chain C has a start vertex and an end vertex, and those two vertices are distinct. So a polygonal chain is a one-dimensional object like a piece of string.

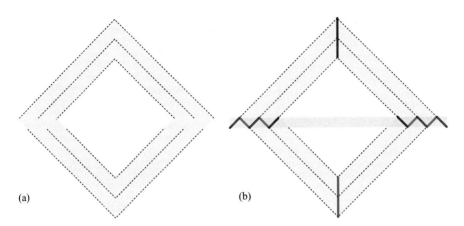

(a)   (b)

Figure 4.4 (a) Nested diamond frames are connected. (b) Uncut strip containing zigzag mountain/valley folds (O'Rourke, 2011, Fig. 3.13). [Reprinted by permission of Cambridge University Press.]

Linkages may have various restrictions on allowable joint movements. Here we will discuss two, illustrated in Fig. 4.5. A *universal joint* allows complete rotational freedom to the two links sharing the joint (although the links cannot coincide, which is interpreted as a form of collision). A *revolute* joint maintains some fixed angle α between the links sharing the joint. The motion permitted by a revolute joint is sometimes called *dihedral motion*. See Fig. 4.6 (and recall Box 1.1 in Chapter 1 for "dihedral"). The chains relevant to the pop-up spinner have a fixed angle $\alpha = 90°$ at each joint, as illustrated in Fig. 4.6(b). Specializing further, we will focus on chains all of whose links have the same length. These are called *unit 90°-chains*, "unit" because we can take the common length to be 1.

4.2.1 Protein Folding

The study of the 3D configurations of such polygonal chains is motivated by protein folding. In fact, a unit fixed-angle polygonal chain can serve as a model for a protein backbone, as illustrated in Fig. 4.7. The actual bond angles along a protein backbone lie roughly in the range of 109° to 122°, so the joints are approximations of a fixed angle α. The actual bond lengths between atoms along the backbone of a protein vary a bit, but not much, roughly between 1.33 Å and 1.52 Å. So again, not quite a unit chain, but approximately the same length.

Biochemists have long been interested in the statistical distribution of the end-to-end lengths of polymers, a class of chain-like molecules that includes proteins and plastics. And to understand the distribution, they need to know the maximum possible end-to-end distance, known as the *maximum span*, or *maxspan* for short, of the chain.

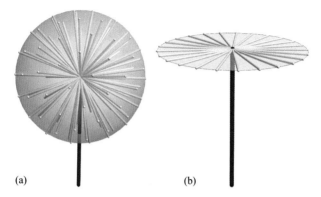

Figure 4.5 (a) Universal joint. With the first link (blue) fixed, the second can reach any point on the sphere (except the south pole). (b) Revolute joint, $\alpha = 90°$. The second link can reach any point on the boundary of a disk perpendicular to the first link.

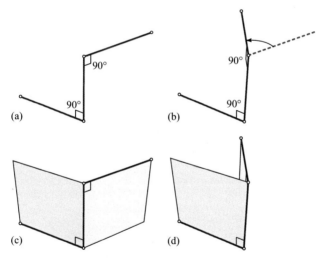

Figure 4.6 Dihedral motion of a 3-link 90°-chain. The 90° angle between adjacent links remains fixed, while the two planes swivel on the shared middle link like a hinged door.

4.2.2 Theorem

Understanding the spinner dynamics hinges on understanding which configurations of a unit 90°-chain achieve the maxspan. The answer is provided by this theorem:

Theorem 4.1 Unit 90°-Chain

The (unique) maxspan configuration of a unit 90°-chain of n links is achieved by the planar staircase configuration.

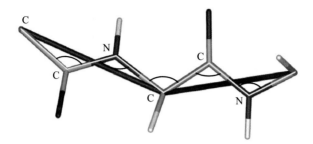

Figure 4.7 Two CCN amino acid cores along the backbone (red), with most side-chain atoms suppressed. The chain and bond angles are superimposed (O'Rourke, 2011, Fig. 3.3). [Reprinted by permission of Cambridge University Press.]

Fig. 4.8 illustrates *staircase configurations* for unit $90°$-chains. They are *planar* in the sense that all the links lie in a common plane.

Perhaps this theorem seems obvious, but even removing the unit-length assumption while retaining the fixed $90°$-angle requirement leads to the maxspan being achieved by a fundamentally 3D configuration, consisting of twisted sections aligned along the line determined by the two end vertices of the chain. Fig. 4.9 shows an example, an 11-link $90°$-chain in a maxspan configuration, whose link lengths vary from 1 to 10. With its vertices labeled $(v_0, v_1, \ldots, v_{11})$, the link's maxspan is the segment $s = v_0 v_{11}$, in this case of length about 39.35. The vertices v_3 and v_8 lie directly on s, partitioning the chain into three planar sections. That the sections are planar is more evident in the overhead view shown in Fig. 4.9(b): the two end sections $\{v_0, v_1, v_2, v_3\}$ and $\{v_8, v_9, v_{10}, v_{11}\}$ lie in a plane parallel to the xy-plane, and the middle section $\{v_3, v_4, v_5, v_6, v_7, v_8\}$ lies in a vertical plane parallel the the z-axis. The staircase configuration of this chain has a slightly smaller span, $37.7 < 39.4$.

> ### Exercise 4.1 Practice: Planar 3-Link Maxspan
>
> What is the maxspan of the 3-link $90°$-chain with link lengths $(1, 2, 3)$, if all three links lie in the same plane?

> ### Exercise 4.2 Challenge: 3-Link Maxspan in 3D
>
> Can you increase the span of the $(1, 2, 3)$ chain from the previous exercise by rotating the third link out of the plane containing the first two links?

So we have to imagine that, say, a 10-link unit $90°$-chain in maxspan configuration might look like Fig. 4.10. Theorem 4.1 says that this configuration does not in fact achieve the maxspan, that the staircase in Fig. 4.8(a) is longer. We now turn to proving the theorem, but only under the very specific assumptions

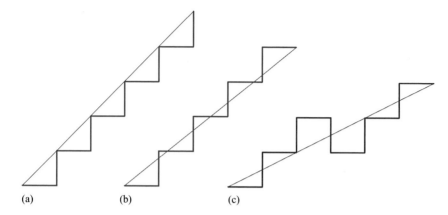

Figure 4.8 Planar staircases of (a) $n = 10$ and (b) $n = 9$ links. (c) Not a staircase: the turns do not alternate left/right.

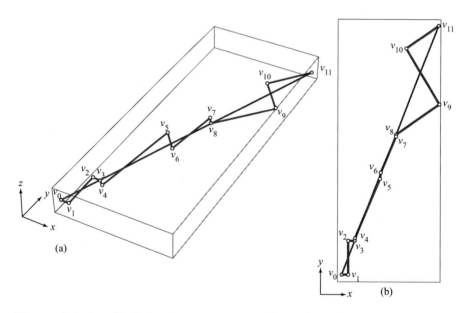

Figure 4.9 An 11-chain in maxspan configuration. Its link lengths are $(1, 5, 1, 1, 10, 3, 6, 1, 8, 10, 6)$, and its span is 39.4. (a) Oblique view; (b) Overhead view (O'Rourke, 2011, Fig. 3.4). [Reprinted by permission of Cambridge University Press.]

of (a) unit length, (b) 90° fixed angle, and (c) an even number n of links. Later we will address generalizations.

We partition the proof into three steps.

(1) Every two adjacent links of a unit 90°-chain form a 45°-45°-90° triangle spanning $\sqrt{2}$ along the hypotenuse, independent of how the two links are

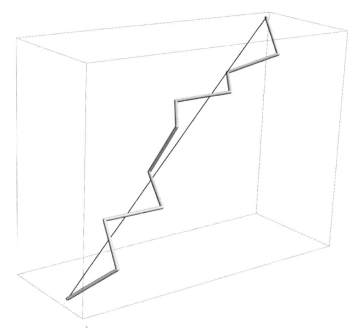

Figure 4.10 A unit 90°-chain of $n = 10$ links, <u>not</u> in maxspan configuration.

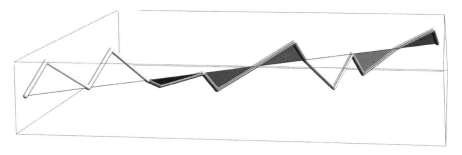

Figure 4.11 Hypotenuse alignment, violating the 90°-angle requirement.

arranged in space. This implies that the largest the maxspan could be is $(n/2)\sqrt{2} = n/\sqrt{2}$, if all the 2-link chain hypotenuses were aligned. (Recall that we are assuming n is even, so $n/2$ is an integer.) But in general this alignment does not respect the 90°-angle requirement at the joints between adjacent pairs of 2-links, as evident in Fig. 4.11.

(2) When can an adjacent pair of 2-links achieve both hypotenuse alignment and 90° at their shared joint v? Fix the first 2-link chain, and imagine spinning the second 2-link chain about v while maintaining 90° at v. As Fig. 4.12 shows, only one position of the second chain continues the first chain's hypotenuse straight and so achieves alignment. And that position places the second chain in the plane determined by the first chain. So the two 2-link chains form a 4-link planar staircase. Repeating the argument,

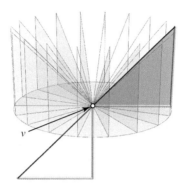

Figure 4.12 Two 2-link chains sharing joint v. Alignment position indicated.

we conclude that the dual requirements of (a) hypotenuse alignment of 2-link chains and (b) 90° joint angles between the 2-link chains are uniquely achieved by the planar staircase configuration.

(3) From step (1) we know that maxspan $\leq n/\sqrt{2}$, and from step (2) we know that the planar staircase uniquely achieves this maxspan. So there is no need to look for a longer configuration: we have proved Theorem 4.1 in the even-n case.

For odd n, the maxspan pierces the intermediate links, as in Fig. 4.8(b), and a proof needs a somewhat different argument, not presented here. (We only need the even-n case for the spinner.)

Exercise 4.3 Practice: Odd-n Maxspan

What is the span of a planar unit-90° staircase when n is odd, as in Fig. 4.8(b)?

Perhaps surprisingly, Theorem 4.1 generalizes to unit fixed-angle chains for any revolute joint angle α: the planar staircase configuration uniquely achieves the maxspan. See Fig. 4.13. This more general theorem more directly applies to protein backbones with $\alpha \approx 115°$. It also requires a rather different proof. So it is when the chain links are not all the same length, as in Fig. 4.9, that 3D configurations achieve the maxspan.

Now we return to the pop-up spinner.

4.3 Unit 90°-Chain in Spinner

Here is the key observation: the mountain/valley zigzag path illustrated in Fig. 4.4 is a unit 90°-chain! It is a chain of creases in the horizontal strip, with the angles fixed to 90° by the construction. When the card is closed, this chain (call it C) is curled up into a spiral like a coiled spring, as crudely depicted

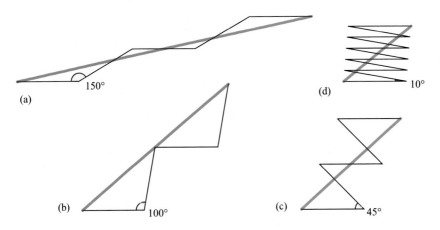

Figure 4.13 Unit α-chains.

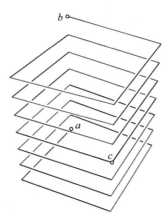

Figure 4.14 The zigzag chain of mountain/valley creases, shown stretched out from a to b in Fig. 4.2(b), curls up when the card is closed, forming a compressed spiral (O'Rourke, 2011, Fig. 3.14). [Reprinted by permission of Cambridge University Press.]

in Fig. 4.14. The diamond frames attached to C are twisted from their flat position as many times as the chain spirals around. When the card is opened, the two endpoints of C at the diamond tips a and b get stretched apart, forcing the chain to head toward its maxspan configuration. But we know from the Unit 90°-Chain Theorem that the maxspan configuration is the *planar* staircase configuration. So C <u>must</u> uncurl from its initial, highly nonplanar state to the planar state, unwinding the initial spiral with all its attached diamond frames. Those larger frames nearest a and b unwind the least; those more deeply nested, smaller frames nearest the centerpoint c have the most to unwind, and so spin faster. The inverse relationship of frame size with spin speed adds to the elegance of the motion.

4.4 Minimal Spinner

Now we see that the unit 90°-chain, curled up when the card is closed, must uncurl to a planar staircase as the endpoints a and b separate to the maxspan of the chain. But this doesn't quite explain how the chain uncurls. To gain a bit more insight into the dynamics of the spinner, we study the "minimal" functioning pop-up spinner. The design is shown in Fig. 4.15. The unit 90°-chain consists of just four links. Point c' is the innermost tip of the spinner, where two cuts meet on the card centerline. We now explore the 3D motion of c' as the card opens, with respect to the points a' and b', which remain fixed to the front and back of the card. See Fig. 4.16.

We concentrate on the midpoint of the chain c and the point c' illustrated, with cc' being the diagonal of a square. It should be clear that throughout the card-opening motion, c and c' remain in the medial plane M. When the card is nearly closed, the chain endpoints a and b are close to one another, and c' is tucked inside between a and b. All three points $\{a, b, c'\}$ are co-located when $\theta = 0°$. As the card opens, c' swings in M on a circle centered on c (which is itself moving in M), increasing the dihedral angle along cc' until it reaches $180°$ when the card is fully opened. It is that swinging motion that drives the spinner's rotation. With this 4-link chain, c' traces out a semicircle. With a longer chain, c' rotates over a number of full circles as the chain unwinds. One can experience this with a physical model, via construction systems such as K'NEX, whose rods and connectors permit easy construction of unit 90°-chains.

Exercise 4.4 Understanding: Minimal Spinner

We emphasized that point c' in the minimal spanner (Fig. 4.16) rotates a full semicircle—$180°$. The central point c also moves on a circle lying in the medial plane M, and in fact on the same circle, call it C. Address these two questions:

(1) Is the radius of C fixed throughout the opening motion?

(2) What is the angular length of the arc of the circle that c traverses?

Finally, we mention that Diana Davis has posted a video (Davis, 2020) explaining how to make a variety of (simpler) pop-up spinners, all based on the same core mechanism of uncurling a unit fixed-angle chain.

Notes

Portions of this chapter are drawn from (O'Rourke, 2011, Ch. 3). There a stronger version of Theorem 4.1 (including odd-n) is proved by induction. The undergraduate thesis that first proved the equivalent of Theorem 4.1 for unit fixed-angle chains is by Nadia Benbernou (Benbernou, 2006; Benbernou

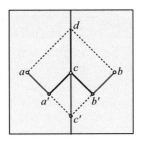

Figure 4.15 Minimal pop-up spinner, with innermost spinner tip c' identified (O'Rourke, 2011, Fig. 3.15). [Reprinted by permission of Cambridge University Press.]

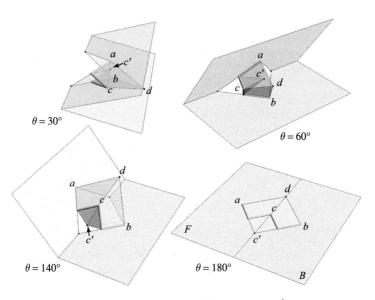

Figure 4.16 Opening progression of minimal spinner: c' swings 180° over the full range for θ. Animation available (O'Rourke, 2021).

and O'Rourke, 2006). Her motivation was protein folding, and it was only after she graduated that I realized her theorem applied to the pop-up spinner. Subsequently she wrote her Ph.D. thesis on related topics (Benbernou, 2011).

5

Polyhedra: Rigid Origami and Flattening

5.1   Polyhedra

For those who love geometry, it would be satisfying to have five pop-up cards, each of which pops up one of the five Platonic solids, shown in Fig. 5.1. Ideally the card would open to $\theta = 180°$ leaving the Platonic solid upright and centered, but achieving this opening to $\theta = 90°$ would also be satisfying. Alas, as far as we know, this has not been achieved. In this chapter, we explore the broader goal of popping up polyhedra, and describe some of the rich (and sometimes "advanced") mathematics surrounding this topic. In the end, we show both 180° and 90° pop-ups of the cube, a 90° pop-up of the octahedron, and a 180° pop-up of the tetrahedron (Sections 5.4, 5.5, and 5.6), but not all of these are entirely satisfactory models. And we are unaware of pop-ups for the dodecahedron or the icosahedron. There is considerable room here for reader ingenuity.

5.1.1   The Five Platonic solids

The five Platonic solids are the most symmetric, the most "regular," of all polyhedra, known for at least 2,500 years. Plato included them in his philosophy,

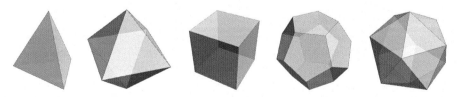

Figure 5.1 The five Platonic solids: tetrahedron, octahedron, cube, dodecahedron, and icosahedron.

and Euclid described them mathematically. Their names derive from their Greek origins: *-hedron* means face or side, while *tetra-*, *octa-*, *dodeca-*, and *icosa-* mean 4, 8, 12, and 20, respectively. The cube is technically a regular hexahedron, having 6 faces.

We now embark on a bit of a detour exploring polyhedra and some of the beautiful mathematics related to their (lack of) flexibility, returning to specific pop-ups in Section 5.4. Chapter 6 will continue the exploration of polyhedra in Section 6.2.

A *polyhedron* is a closed 3D shape composed of flat, polygonal faces joined edge to edge. Each *edge* of a face is a line segment shared by exactly two faces. A polyhedron's corners are called *vertices* (singular: *vertex*). We define a polyhedron as a surface, but one should imagine that the surface bounds a solid: it could be filled with water without leaking. Defining a polyhedron precisely is difficult, but this will suffice for us.

The faces are flat 2D polygons, where a *polygon* is a region of a plane bounded by a non-self-intersecting closed path formed by a sequence of straight-line segments, its *edges*, which meet at *vertices*. Often they are called *simple polygons*, where "simple" emphasizes the lack of self-crossings and having no holes.

A *regular polygon* is one all of whose edges have the same length, and all of whose interior vertex angles are identical. The equilateral triangle has three 60° angles, the square four 90° angles, the regular pentagon five 108° angles, and the regular hexagon six 120° angles. Note that both edge lengths and vertex angles are specified in the definition of "regular": as we saw in Chapter 1, a rhombus is a quadrilateral with all edge lengths equal, but it is not considered a regular polygon.

Finally, the *regular polyhedra* are the polyhedra all of whose faces are the same regular polygon, with the same number of faces incident to each polyhedron vertex, "incident" being the technical term for "touching"—in this case, sharing a vertex. It is an interesting exercise to prove that the five Platonic solids are the only regular polyhedra. Note, for example, that one cannot build a polyhedron solely from regular hexagons, because placing three incident to a vertex creates $3 \times 120° = 360°$ around that vertex, which implies that it is in fact not a corner but instead flat.

Exercise 5.1 Practice: Truncated Icosahedron

The *truncated icosahedron* is a semi-regular polyhedron, better known as a soccer ball. Each vertex is shared by two regular hexagons and one regular pentagon. What is the total angle surrounding (incident to) each vertex?

5.1.2   Convex Polyhedra

The five Platonic solids are special cases of what are known as *convex polyhedra*. Convexity is an important mathematical concept, which can be summarized

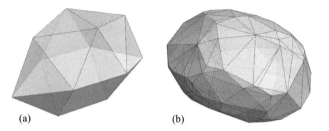

Figure 5.2 (a) Sphenomegacorona, $n = 12$ vertices. (b) Random points on an ellipsoid, $n = 100$ vertices.

informally as having no dents. Technically, a shape P is convex if, for every two points a and b inside P, the line segment ab is wholly inside P. If there were a dent in P, then selecting a and b to straddle the dent would result in some portion of ab lying exterior to P.

One can think of convex polyhedra as cut gems. Fig. 5.2 illustrates two. The shape in (a) is formed of regular polygons (16 equilateral triangles and 2 squares), but some vertices are incident to a square and some are not. This shape is known as a *semi-regular* polyhedron. It's been given an impressive name: the sphenomegacorona. The shape in (b) is an irregular convex polyhedron.

Convex polyhedra possess mathematical properties not shared by all polyhedra, some of which will be relevant in Section 5.3.

5.1.3 Nonconvex Polyhedra

Nonconvex polyhedra can be quite complicated. For example, Fig. 5.3 shows a richly complex nonconvex polyhedron. And we'll see more Chapter 6. We now turn to the challenges presented by pop-up polyhedra.

Exercise 5.2 Challenge: Nonconvex Platonic Solid

Can any of the five Platonic solids be reformed into a nonconvex polyhedron with the exact same surface structure, i.e., the same number of the same regular polygons surrounding each vertex?

Exercise 5.3 Understanding: Cube + Octahedron

View the top half of an octahedron as a pyramid P on a square base. Glue six copies of P onto the faces of a cube. Is the resulting polyhedron convex or nonconvex?

Figure 5.3 Triangulated mesh of the head of Michelangelo's *David* (Valette, Chassery, and Prost, 2008, Fig. 13). [©IEEE. Reprinted by permission of Institute of Electrical and Electronic Engineers.]

5.2   Rigid Origami

The cardstock used in pop-up cards is stiff, unlike the paper used in origami. A relatively new topic in the origami world is what is known as *rigid origami*. Here flat faces are viewed as rigid steel plates, with two faces hinged along shared edges. The plates cannot bend nor pass through one another. This greatly reduces the shapes that can be achieved—the iconic origami crane can not be rigidly folded from its crease pattern; see Fig. 5.4. Origami depends on the ability to bend the paper during construction. To pop-up a polyhedron, the ideal would be to find a rigid-origami design that permits rigid polyhedron faces to hinge on edge-creases to allow motion from the flat state—card closed—to a fully opened polyhedron. We will see shortly that this goal cannot be achieved as just stated, but first we provide a few examples of rigid origami.

5.2.1   Rigid Origami Examples

Square Twist.   The "square twist" is a beautiful mechanism that moves from a flat configuration to a square half the size. It is a well-known design in the

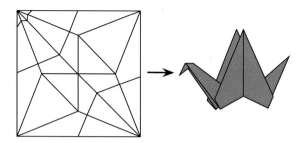

Figure 5.4 Origami crane and crease pattern (Hull, 2020, Fig. 1 (detail)). [Reprinted by permission of Cambridge University Press.]

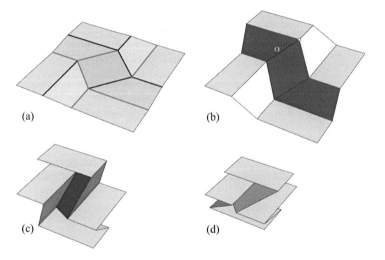

Figure 5.5 The square twist (O'Rourke, 2011, Fig. 6.8). [Reprinted by permission of Cambridge University Press.]

origami world, used, for example, as part of the "Kawasaki rose." Fig. 5.5 illustrates several snaphots. For our purposes, this serves as an example of a *one degree-of-freedom* mechanism, which, as we saw in Chapter 2, means that the motion is controlled by one parameter that varies throughout the motion. For the square twist, that parameter could be taken to be the dihedral angle α, which varies from 180° in the initial state (a), to 0° in the final flattened state (d). This is certainly not obvious, but if you build the design, you can feel in your fingers that there is a single degree of freedom, a single motion path in its "configuration space."

Deployable Solar Arrays. Rigid origami is perfect for compacting solar array panels and unfolding them in space. An example is shown in Figs. 5.6 and 5.7. The inventors call this the *HanaFlex* array. *Hana* is Japanese for "flower," and the design is modeled on the origami "flasher" model invented

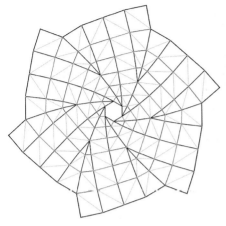

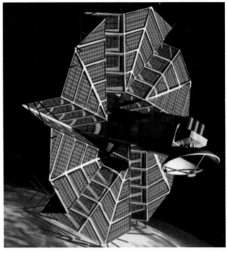

Figure 5.6 Six-sided flasher folding pattern (Zirbel, Lang, Thomson, Sigel, Walkemeyer, Trease, Magleby, and Howell, 2013, Fig.4(b)). [©ASME. Reprinted by permission of American Society of Mechanical Engineers.]

Figure 5.7 HanaFlex solar array. [Snapshot from YouTube video, by permission of authors Larry Howell and Spencer Magleby.]

by Chris Palmer and Jeremy Shafer in 1994. Renowned origamist Robert Lang collaborated with the engineers to design HanaFlex to wrap tightly around the Atlas V rocket during launch, then deploy to a 25-meter array of rigid solar panels in space.

Shopping Bag. My favorite example of rigid origami is the standard grocery shopping bag, seen in Fig. 5.8. Of course it is not built from rigid pieces, but rather thick yet still pliable "bag paper." But if it were built from rigid plates, hinged at its creases, then it has been proved that it has only two possible states: fully opened, as in the figure, or completely flattened. So the opened state is rigid, as there are no intermediate states to reach its flattened configuration. And the reverse holds: it is rigidly stuck in its flattened state. This is in fact one of the reasons the shopping bag is such a successful engineering design: because its faces are stiff (but not rigid), it tends to stay open rather than spontaneously collapsing, because collapsing requires bending the resistant bag paper.

Exercise 5.4 Practice: Rigid Cube

A cube is rigid. Does it remain rigid

- with the top face removed?
- with two adjacent faces removed, say, top and front faces?
- with both the top and bottom faces removed?

Figure 5.8 A grocery shopping bag (O'Rourke, 2011, Fig. 6.1). [Reprinted by permission of Cambridge University Press.]

5.2.2 Rigid and Flexible Polyhedra

Now that we have defined polyhedra and the rigid origami model, we turn to exploring the rigidity and flexibility of polyhedra.

Cauchy Rigidity. Augustin Cauchy is known for his many contributions to mathematics, including the introduction of $\varepsilon - \delta$ limit arguments to calculus. In the early nineteenth century, he proved what is now known as Cauchy's Rigidity Theorem: convex polyhedra are rigid. If all the faces of a convex polyhedron P are rigid—flat plates—then even if every edge of P is a hinge, P is rigid. As we've seen, the same statement for convex polygons in the plane is false: for example, a square with rigid bars as edges and universal joints at its corners flexes to rhombi. The rigidity of convex polyhedra is by no means obvious, and Cauchy's proof is clever and intricate. In fact his proof contained a flaw that was only uncovered and corrected a century later.

So this theorem rules out popping up the Platonic solids, or any convex polyhedron, in the ideal case. Before we turn to ways to circumvent this basic theorem, we continue to explore the flexibility of polyhedra.

Connelly's Flexible Polyhedron. A natural question following Cauchy's Rigidity Theorem is: are there any flexible <u>nonconvex</u> polyhedra? Indeed, Euler conjectured that all polyhedra are rigid even before Cauchy found his proof. This was an open question for over a century until Robert Connelly constructed a flexible nonconvex polyhedron in 1978. His construction has subsequently been simplified, but all the known flexible nonconvex polyhedra flex only slightly, distorting just a small amount.

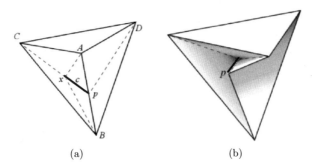

(a)   (b)

Figure 5.9 Rigid flattening of a slit tetrahedron. The slit c connects x to p. p (Abel, Connelly, Demaine, Demaine, Hull, Lubiw, and Tachi, 2015, Fig. 1). [Reprinted by permission of American Mathematics Society and Cambridge University Press.]

Bellows Theorem. There is an interesting constraint behind this small-flexing. A remarkable result known as the Bellows Theorem says that any such rigid-origami polyhedron has a constant, fixed volume as it flexes. The reason it is called the "Bellows" Theorem is that it implies that there cannot be a polyhedral bellows, where flexing expels a volume of air through a hole. To enable a polyhedron to substantially flex, there must be cuts or slices along the boundary—for leaving it a closed polyhedron with rigid faces ensures that it could never flatten to zero volume.

Rigid Flattening via Slits. A recent study introduced a notion of *rigid flattening* of a polyhedron after the addition of a number of cuts, or *slits*, and possibly additional hinged creases. The slits should not disconnect the surface. The authors also insisted that all the slits close up completely in the flattened state, not a condition needed for pop-up design, but still a natural constraint. They showed that the regular tetrahedron of unit edge-length could be rigidly flattened after adding four creases and a tiny slit of length only 0.05. The flattening motion has a single degree of freedom. An easier-to-see example uses a longer cut, of length $1/(2\sqrt{3}) \approx 0.29$: see Fig. 5.9.

Much remains unknown on this mathematical topic. And in particular, it is unknown whether similar crease-and-slit alterations would allow rigid flattening of any given convex polyhedron.

When mathematicians isolate a clearly defined problem with as-yet unknown resolution, they designate it as an *open problem*, and advertise it to the community in the hope that someone will settle it, a practice we will follow in this book. Here is our first open problem:

> **Open Problem 5.1 Rigid Flattening of Convex Polyhedra**
>
> Can every convex polyhedron be rigidly flattened using slits that do not disconnect the surface and adding at most a finite number of hinged creases?

We now turn to a second method of circumventing the inflexibility of polyhedra.

5.3 Continuous Flattening

We all regularly flatten cereal boxes for recylcling by simply stomping on them. Less frequently do we flatten a closed box, but it is not difficult to do so by denting the four sides, as illustrated in Fig. 5.10. However, this flattening cannot be achieved by rigid flattening if the box is unslit, as we have seen. A concept was introduced in 2001 to capture such flattening: *continuous flattening*. The idea is to allow "moving" or "rolling" creases, allowing the faces of the polyhedron to bend, but not tear or pass through one another. Although this model of flattening takes us beyond what might help in pop-up design, it is nevertheless closely related mathematically to rigid flattening.

Among the first substantive results on continuous flattening was a method to flatten each of the five Platonic solids. We illustrate the idea for the cube in Fig. 5.11.

Label the vertices of the cube v_1, v_2, v_3, v_4 on the bottom face, and v_5, v_6, v_7, v_8 around the top face, as in Fig. 5.11(a), with a the midpoint of $v_1 v_5$ and c the midpoint of the front face. Let h be the height of the cube, initially $h = 1$ and ultimately $h = 0$ when completely flattened. The right triangle $\triangle v_1 c v_5$ will undergo rolling creases, depicted by the closely spaced blue lines. As h decreases, the ac segment buckles at point b, forming four $\triangle$:

$$\triangle abv_1, \ \triangle bcv_1, \ \triangle abv_5, \ \triangle bcv_5 \ .$$

Point b slides along the crease incident to a, eventually ending up under the center of the top when completely flattened. Because the length of ac is $\frac{1}{2}$, we know that the sum of the distances from b to a and to c is this distance:

$$|ab| + |bc| = \frac{1}{2} \ .$$

Thus b must lie on a horizontal ellipse with foci a and c: the sum of the distances from the moving point b from the fixed points a and c is a constant. This, together with the constraint that b lies on the crease incident to a, determines b's position.

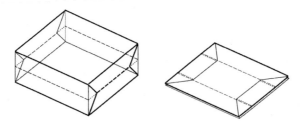

Figure 5.10 Flattening a box by denting the sides (Demaine and O'Rourke, 2007, Fig. 18.2). [Reprinted by permission of Cambridge University Press.]

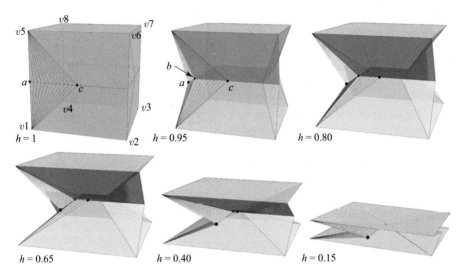

Figure 5.11 Continuous flattening of a cube. The red point b moves with the decrease of height h. Animation available (O'Rourke, 2021).

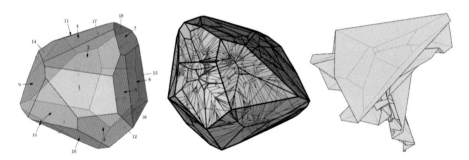

Figure 5.12 Continuous flattening of a convex polyhedron. Left: face ordering for "orderly squashing." Center: Crease pattern. Right: Final flat state, with the back face on top (Abel, Demaine, Demaine, Itoh, Lubiw, Nara, and O'Rourke, 2014, Fig. 3). [Reprinted by permission of the authors.]

Although only the triangle $\triangle v_1 c v_5$ is marked in the figure with rolling creases, symmetric triangles on each of the four side faces are similarly creased. Note that all other faces are rigid. In particular, each side face consists of two rigid quadrilaterals hinged on their shared horizontal edge. So 5/6 of the surface consists of rigid faces: 3/4 of each side face, plus the top and bottom faces. The motion is again a single degree-of-freedom motion, controlled by the parameter h. Similar continuous flattenings have been achieved for all five Platonic solids. However, their techniques were not sufficiently powerful to extend to all convex polyhedra.

That goal—the continuous flattening of any given convex polyhedron—was achieved in 2014 via an algorithm whose intricacy is evident from Fig. 5.12.

Still, it left open whether this could be achieved for all polyhedra, both convex and nonconvex.

This was finally resolved in 2020. However, at any one time, the flattening model used to achieve this result may have an infinite number of creases, unlike the cube example in Fig. 5.11: that example has 28 creases at any one time: 16 moving creases (4 per side face) and 12 fixed creases. So, if we insist on a finite number of creases, the long pursuit to continuously flatten any polyhedron is not yet over.

5.4 Pop-Up Cubes

To review, to pop up a polyhedron, its faces should be rigid, as is cardstock, but it needs to have slits and possibly extra fixed creases to enable the model to flatten. There is the added challenge of arranging for the popping-up motion to be mechanically driven by the card opening. The simplest polyhedron to pop up is the cube. We describe two different constructions, one that pops up a cube when the card is fully opened to $\theta = 180°$, and one that instead aims for a popped-up cube at $\theta = 90°$.

5.4.1 Cube at 180°

This first construction is especially easy to understand. Let the cube be a unit-cube in the sense that all its edges have length 1. Label the vertices of the cube as we did in Fig. 5.11: the cube base is v_1, v_2, v_3, v_4, and the cube top is v_5, v_6, v_7, v_8. Think of the face v_1, v_2, v_6, v_5 as the front face of the cube, attached to the card back B along $v_1 v_2$, as shown in the nearly fully opened ($\theta = 150°$) configuration in Fig. 5.13 (left). There is no cube bottom, and the top edges $v_5 v_8$ and $v_6 v_7$ are slit open. The top, the left, and the right cube faces are creased down the middle, each into two rectangles. The top crease is a valley, while the side creases are mountains. So the top folds inside the cube

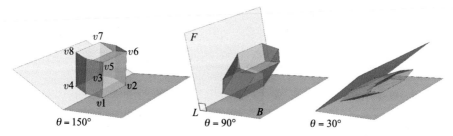

Figure 5.13 Cube pop-up at three θ angles. Animation available (O'Rourke, 2021).

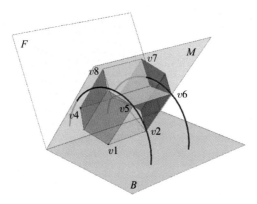

Figure 5.14 Vertices v_5 and v_6 travel on unit-radius circles. Animation available (O'Rourke, 2021).

as the card is closed, while the sides bow outward, avoiding intersection with the top.

> ## Exercise 5.5 Understanding: Flattened Cube Area
>
> For a unit cube, the completely flattened shape (Fig. 5.13, right) covers an area of 2: One square face, and two $1 \times \frac{1}{2}$ side rectangles. While retaining the final cube pop-up at $\theta = 180°$, can the model be modified to reduce the $\theta = 0°$ area coverage?

The geometry of the motion is especially simple. The top vertices of the front face v_5 and v_6 each ride on unit circles centered on v_1 and v_2, respectively, as shown in Fig. 5.14. The creases in both the top and the side faces lie in the medial plane M throughout the motion. The back cube face is parallel to the front cube face. The front cube face makes an angle of $\theta/2$ along the segment $v_1 v_2$ where it meets the card back B, because the front cube face is parallel to M. Finally, the dihedral angles δ at the top and side creases are all equal and exactly θ. Let us prove this dihedral-angles claim.

From a side view, with line of sight parallel to the card centerline L, one can see a parallelogram as overlayed in Fig. 5.15, which shows that the half angle along the valley crease in the cube top is $\theta/2$. So the dihedral angle along that crease is θ. Thus the front half of the top is parallel to B and the back half of the top is parallel to F.

It is a bit more difficult to see the angle along the side mountain creases, although the reasoning is similar to our analysis of V-folds in Section 2.1.4. Fig. 5.16 removes several faces and adds labels so we can focus on the details. The triangle $\triangle v_1 v_4 a$ has the centerline angle θ incident to a. The triangle $\triangle v_1 v_4 b$ determines the dihedral angle δ along the crease bc. The two triangles are congruent isosceles triangles because (a) they share the edge $v_1 v_4$ and (b) the other edges are each of length $\frac{1}{2}$. Therefore $\delta = \theta$.

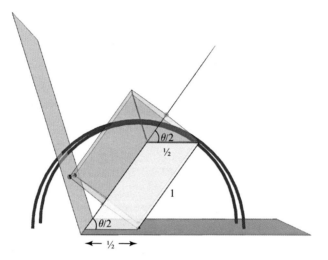

Figure 5.15 For a unit cube, the (yellow) parallelogram is $1 \times \frac{1}{2}$.

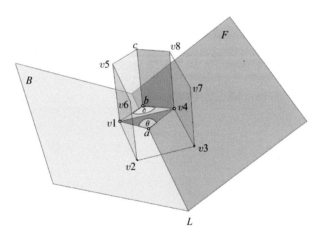

Figure 5.16 Triangles $\triangle v_1 v_4 a$ and $\triangle v_1 v_4 b$ are congruent, and so $\delta = \theta$.

Even though this almost doesn't deserve "theoremhood," for consistency with Theorem 2.1 (and Theorems 5.2 and 5.3 to come), we record here the dihedral-angle relationships just detailed:

Theorem 5.1 Cube $\delta = \theta$

The three creases (top and sides) in the pop-up cube model that lie in the medial plane M (Fig. 5.13) each have dihedral angle $\delta = \theta$.

5.4.2 Cube at 90°

There are several possible constructions that achieve a popped-up cube at $\theta = 90°$, with one face of the cube attached to the front face F of the card, which then drives the opening. However, these constructions are not entirely satisfying, because it seems difficult to make them work structurally without adding string or a rubber band to fully close the cube. Nevertheless, we describe one with attractive 3D geometry, whose mechanism will play a role in an algorithm discussed in Chapter 6 (Fig. 6.19). And we challenge readers to improve the construction.

The construction and mechanism are illustrated in Fig. 5.17. Let us say that the pop-up is again a unit cube, labeled as before: at $\theta = 90°$, v_1, v_2, v_3, v_4 are fixed on the card back B, and v_5, v_6, v_7, v_8 are the cube top, with v_7, v_8 attached to the card front F. The side panel v_1, v_5, v_8, v_4 is creased along $v_4 v_5$, and detached from the top and front edges by slitting edges $v_1 v_5$ and $v_5 v_8$. The same construction is repeated on the other side panel, v_2, v_6, v_7, v_3.

As θ is reduced from 90° to fully closed at 0°, the top and front faces of the cube move rigidly as a rhombus, just as in Fig. 1.6 in Chapter 1. Meanwhile vertex v_5 "splits" into v_5 moving with the rhombus and what is labeled v_5' in the figure. We now analyze the motion of v_5' and its two incident triangles.

Exercise 5.6 Understanding: Cube 90°

What is the total distance in space traveled by vertices v_5 and v_5' during the opening of the card from 0° to 90° in Fig. 5.17?

Both triangles $\triangle v_4 v_1 v_5'$ and $\triangle v_4 v_8 v_5'$ remain rigid throughout the motion. Focusing on triangle $\triangle v_4 v_1 v_5'$, its base edge $v_4 v_1$ remains fixed on B, while v_5' rotates down, eventually to also lie on B at $\theta = 0$. Thus v_5' rides on a fixed, vertical, unit-radius circle centered on v_1. This is illustrated in Fig. 5.18. Similar reasoning leads to the conclusion that v_5' also lies on a horizontal unit circle centered on v_8. However, the complication is that v_8 moves as a function of θ, whereas v_1 is fixed to B. The key observation is not that this v_8 circle is horizontal—it is only horizontal at $\theta = 90°$—but rather the circle is orthogonal to the edge $v_4 v_8$, about which v_5' swivels. This is most evident from the 50° snapshot in the figure.

How can we compute the location of v_5' at the intersection of the two circles? We know that v_5' is at all times a distance $\sqrt{2}$ from v_4 (along the diagonal), and a distance 1 from v_1 and v_8:

$$|v_4 v_5'| = \sqrt{2}\,, \quad |v_1 v_5'| = |v_8 v_5'| = 1\,.$$

Vertex v_8 is moving, but we know its relationship to θ. So we could view v_5' as sitting at the intersection of three spheres, S_1 and S_4 fixed and S_8 moving, where the indices refer to the vertex labels of the centers. And as we've seen in Section 2.1.2 (Fig. 2.5), in general three spheres intersect in two points.

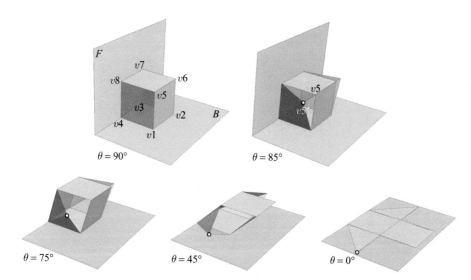

Figure 5.17 Cube pop-up, with edges slit so that the side panels (vertex v_5' highlighted) can bow outward. The front face F of the card is removed in the last three snapshots. Animation available (O'Rourke, 2021).

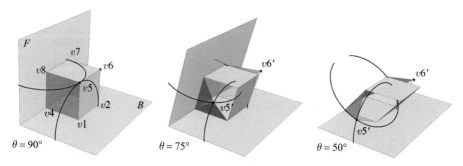

Figure 5.18 Vertex v_5' rides on two circles, one fixed centered on v_1 and one tilting centered on v_8.

Consider another viewpoint: two spheres intersect in a circle, so the intersection of three spheres A, B, C can be viewed as the intersection of the circle $A \cap B$ with the circle $B \cap C$. In our case, the two unit-radius circles are each the intersection of unit-radius spheres S_1 and S_8, respectively, with the $\sqrt{2}$-radius sphere S_4 centered on v_4. See Fig. 5.19.

We offer a final remark. One might hope, just for aesthetic reasons, that the dihedral angle δ along the diagonal crease $v_4 v_5'$ is some straightforward function of θ, as we found for the V-fold and for the 180° cube (Theorems 2.1 and 5.1). At $\theta = 90°$, δ is 180°: the side face is flat. At $\theta = 0°$, δ is 0°, as it is double-folded flat along the crease. So $\delta = 2\theta$ is a feasible relationship, matching at the endpoints. However, the relationship is more subtle, as shown in Fig. 5.20: δ is

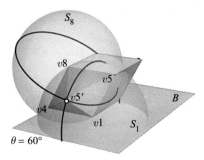

Figure 5.19 The vertical circle is $S_1 \cap S_4$, and the tilted circle is $S_8 \cap S_4$. (Sphere S_4 is not shown.) The two circles meet at v_5' (and at v_6').

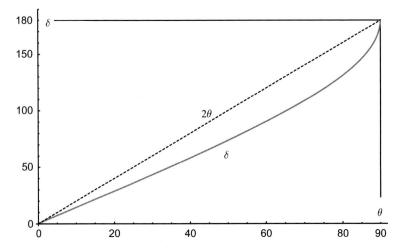

Figure 5.20 Computed relationship between the card angle θ and the dihedral angle δ at the side-face $v_4 v_5'$ crease.

a bit smaller than 2θ except at the extremes. In Section 5.6 we will encounter another "nice" dihedral angle relationship.

An unsatisfying aspect of the cube pop-up in Fig. 5.17 is that one finds it difficult to arrange for the side panels to completely close and seal the cuts along the top and front edges. No doubt a clever designer could add some internal structures that effectively pull those panels tightly closed at $\theta = 90°$.

5.5 Pop-Up Octahedron at 90°

An octahedron can be viewed as back-to-back pyramids sharing a square base Q. One can build an octahedron pop-up with one 90° corner of Q on the card centerline L, two edges of Q on the card back B and front F, and—unusually— nothing else glued to B or F besides those edges. (That just edges ab and ac are glued to B and F creates some construction challenges.)

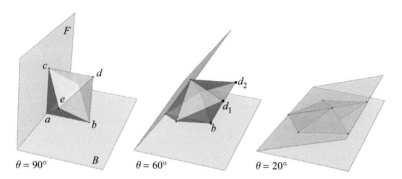

Figure 5.21 Octahedron pop-up at several θ values. Animation available (O'Rourke, 2021).

Let us fix all edges of the octahedron to be unit length; so Q is a unit square. Label the six vertices of the octahedron as illustrated in Fig. 5.21: $Q = abdc$, and e and f are the apexes of the pyramids. All eight triangles are equilateral. We slit two edges of Q, cd and bd; there is no need to add creases beyond the octahedron edges. At $\theta = 90°$, the octahedron is fully popped up. As θ reduces toward $0°$, vertex d splits into d_1 and d_2. At $\theta = 0°$, the angle $\angle d_1 b d_2$ reaches $120°$, filling in the gap left by the four equilateral triangles incident to b.

One can see that vertex e lies at the intersection of three spheres, each of radius 1: the static S_b and S_a, and the moving S_c (moving with F). And d_1 also lies at the intersection of three unit spheres: the static S_b, and the moving S_c and S_e. The resulting motion is quite elegant, with many pleasing symmetries.

There are a total of 10 dihedral angles: an octahedron has 12 edges, but we slit open two. Although none of the dihedrals seem to have a simple relationship to θ, the various symmetries result in only two numerically distinct dihedral angles among the 10. We now prove this claim.

Let δ_{ba} be the dihedral angle along the edge ba, and use similar notation for the other dihedrals. Let $\omega \approx 109°$ be the dihedral of every edge at $\theta = 90°$. Then:

Theorem 5.2 Octahedron Dihedrals

Of the 10 dihedrals, the four lying in the midplane M are small ($\leq \omega$) and equal:
$$\delta_{ea} = \delta_{ed_1} = \delta_{fa} = \delta_{fd_2} \ .$$
The other six are large ($\geq \omega$) and equal:
$$\delta_{ba} = \delta_{be} = \delta_{bf} = \delta_{ca} = \delta_{ce} = \delta_{cf} \ .$$

Several of the equalities in this theorem follow by noting symmetries: The mechanism is symmetric left and right, and also top and bottom. Just one equality among the four small angles, and one equality among the six large, will determine all the others. We focus on proving equality of two large angles,

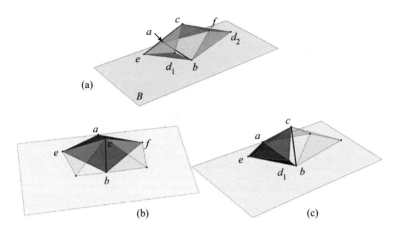

(a)

(b)   (c)

Figure 5.22 (a) Front face F removed, and two equal obtuse edges ba and ce highlighted. (b) Four triangles incident to a. (c) Four triangles incident to e.

$\delta_{ba} = \delta_{ce}$, in three steps. It may help to remember that reversing the indices leaves the angle the same: $\delta_{ba} = \delta_{ab}$ and $\delta_{ce} = \delta_{ec}$. Fig. 5.22(a) highlights the edges ab and ec.

(1) Consider just two equilateral triangles joined along an edge, say, $\triangle abe$ and $\triangle abf$ sharing edge ab. The dihedral angle δ_{ab} along ab is determined by the distance between the two triangle tips, $|ef|$. In fact, $\delta_{ab} = 2\sin^{-1}(|ef|/\sqrt{3})$, but we only need that the angle is determined, not its actual value.

(2) Now we study four equilateral triangles connected in a cycle, such as the four triangles incident to a vertex of the octahedron. In particular, we'll use the four triangles incident to a, shaded in Fig. 5.22(b):

$$\triangle abe \,, \triangle ace \,, \triangle abf \,\, \triangle acf \,.$$

We make two claims: the four dihedrals are determined by the distance $|bc|$, and the dihedrals come in equal pairs. Viewing the four triangles as two pairs of two triangles, we can see that δ_{ae} is determined by $|bc|$, and δ_{af} is also determined by $|bc|$. So $\delta_{ae} = \delta_{af}$. Similar reasoning using the distance $|ef|$ allows us to conclude that $\delta_{ab} = \delta_{ac}$.

(3) Next we view the four triangles as a part of the full octahedron. We have just seen that δ_{ab} is determined by $|bc|$. Look at the four triangles shaded in Fig. 5.22(c):

$$\triangle abe \,, \triangle ace \,, \triangle bed_1 \,\, \triangle ced_1 \,.$$

The distance $|bc|$ plays the same role for these four triangles as it did for the four in Fig. 5.22(b). So we can conclude that $\delta_{ab} = \delta_{ce}$, the two highlighted edges in Fig. 5.22(a).

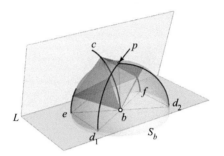

Figure 5.23 Tracks of octahedron vertices on sphere S_b. Animation available (O'Rourke, 2021).

Thus we have established that two of the large dihedrals are equal, and equality with the other four follows by the symmetries of the construction.

A similar argument can show that two of the small dihedrals are equal, say, $\delta_{ea} = \delta_{ed_1}$, and equality with the other two is determined by symmetry. So we have established Theorem 5.2: six equal large dihedrals and four equal small dihedrals.

Exercise 5.7 Understanding: Two Triangle Dihedrals

In step (1) we claimed that $\delta_{ab} = 2\sin^{-1}(|ef|/\sqrt{3})$. Derive this formula.

More insight to the dynamics of the model can be seen from the tracks in space of the vertices during the motion. In particular, fix the unit-radius sphere S_b centered on b. The vertices d_1, d_2, e, and f all lie on S_b, because they are each connected to b by equilateral triangle edges. Starting from $\theta = 0°$, d_1 and d_2 rise up with increasing θ and meet at the north pole p of S_b, where they join to form vertex d at $\theta = 90°$. See Fig. 5.23. Vertices e and f swivel about edge ab, and so their (short) tracks lie in a vertical plane that includes the line through the $\theta = 0°$ positions of those vertices.

The tracks of d_1 and d_2 appear to lie on what are known as *great circles* of S_b: circles whose centers coincides with b. A great circle is the intersection with S_b of a plane through b. In our case, this would be a plane including $\{d_1, b, p\}$, where p is the north pole directly above b. Despite appearances, neither d_1 nor d_2 tracks a great circle of S_b, although the traces are quite close to great circle arcs. It is a challenging exercise to show that, in this instance, appearances are deceiving.

Exercise 5.8 Challenge: Not Great Circles

In Fig. 5.23, it appears that d_1 traces out an arc of a great circle, the intersection of a vertical plane through the center b of the sphere S_b. Prove that this is in fact <u>not</u> the case.

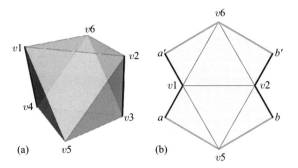

Figure 5.24 Octahedron. Red edges are cut. Brown edges are folds.

5.6   Rubber Band Pop-Ups

We have restricted our pop-up designs to exclude strings or levers or rubber bands, relying solely on the card opening angle θ to drive all the dynamics. It is, however, worth mentioning that there are rubber band–driven pop-ups for all five of the Platonic solids.

Fig. 5.24 illustrates the doubly-covered flat state for such a pop-up for the octahedron. Edges v_1v_4 and v_2v_3 are cut, and the two incident triangles are creased to the midpoints of those edges. So in Fig. 5.24(b), points a and a', and b and b', are images of those midpoints. Stretched rubber bands connect a to b', and b to a'. Then, once flattening pressure is removed, the flat state pops up to a 3D octahedron. Curiously, the most difficult rubber band pop-up is the tetrahedron, which is the easiest card pop-up, to which we now turn.

5.7   Pop-Up Tetrahedron at 180°

In some sense a pop-up regular tetrahedron is like a V-fold in reverse: at $\theta = 180°$, a V-fold is flat, as seen in Figs. 2.1 and 2.2 in Chapter 2. But we want the tetrahedron to be fully popped up when $\theta = 180°$. We will see that there are parallels between the geometry of a V-fold and a 180° pop-up of the regular tetrahedron. We now describe the construction, referring to Fig. 5.25.

Name the four corners of the tetrahedron T a, b, c, d, with d the apex of T and $\triangle abc$ the base. With the card fully opened at $\theta = 180°$, a is on the centerline L of the card, and the line segment bc straddles L, with b on the back face B and c on the front face F. The apex d sits above the plane of the opened card. The base triangle is removed, but the remaining structure is still rigid. Flexibility is introduced by adding a crease down the altitude of equilateral triangle $\triangle bcd$ from the apex d to point p. As before, it is convenient to refer to a point position at a particular value of θ: d_0 refers to the location of apex d when the card is closed at $\theta = 0°$, and d_π and p_π refer to the positions of d and p, respectively, when the card is fully opened at $\theta = 180°$.

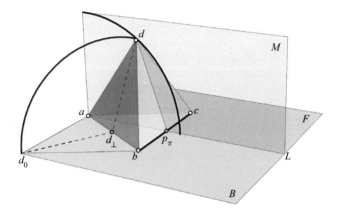

Figure 5.25 Pop-up of the regular tetrahedron $abcd$. Here $\theta = 180°$.

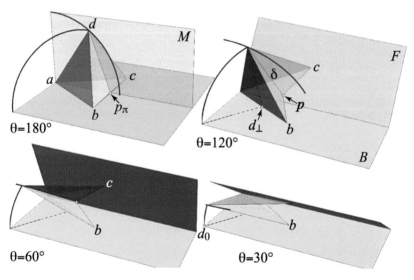

Figure 5.26 The apex d follows both the fixed red arc and the blue arc in the moving medial plane M. Animation available (O'Rourke, 2021).

Let $d_\perp$ be the projection of the apex of T down the face $\triangle abp$ onto the base edge ab, i.e., $d_\perp$ is the foot of $\triangle abd$'s altitude. Then, with the card back fixed to the xy-plane, as the card opens, the apex d follows the circular arc (red in Fig. 5.25) centered on $d_\perp$, from d_0 to its final position d_π. This circle arc is the rim of a cone apexed at a swept out by the rotating segment ad. The motion of d in the medial plane also follows a circular arc (blue in the figure), because the edge ad remains in the medial plane and has fixed length throughout the motion. But, just as we saw with the V-fold (Eq. 2.1), the relationship of these circular arc motions with respect to the card opening angle θ is not straightforward, i.e., not simple functions of θ.

Fig. 5.26 shows a few snapshots of the card closing.

5.7.1   Dihedral Angle $\delta = \theta$

One might hope that the $\delta = \theta$ regularity we noticed in the V-fold along the segment ap (Theorem 2.1) might also hold along the tetrahedron's comparable segment ad. Let us call this dihedral angle, between $\triangle abd$ and $\triangle acd$, δ'. The dihedral angle between any pair of faces of a regular tetrahedron is known to be $\cos^{-1} \frac{1}{3} \approx 71°$. So the relationship between θ and δ' cannot be so simple, as it must show that $\theta = 180°$ leads to this $\delta' \approx 71°$, because we achieve a regular tetrahedron when the card is fully opened.

However, there is a nice $\delta = \theta$ regularity, but along dp rather than ap. Call the dihedral angle along the valley-fold edge dp again δ. It is exactly equal to the card angle θ, for any tetrahedron T symmetric across the card centerline, symmetric in the sense that $|bp| = |cp|$. So this relationship holds a bit more generally than just for a regular tetrahedron.

Theorem 5.3 Tetrahedron $\delta = \theta$

Using the presented notation, for any tetrahedron symmetric about the card centerline L, the dihedral angle δ along the dp valley crease is exactly the centerline angle θ throughout the motion.

We now proceed to prove that $\delta = \theta$ in a manner similar to the $\delta = \theta$ proof for the V-fold in Section 2.1.4. Refer to Fig. 5.27 throughout.

The proof consists of six claims. Let's establish the coordinate system as follows: a is the origin, the card back is fixed in the xy-plane, and the centerline L lies along the y-axis.

(1) The triangle $\triangle bp_\pi c$ is orthogonal to L throughout the motion. This is because b, p_π, c all have the same fixed y-coordinate. Therefore, the plane containing $\triangle bp_\pi c$ is orthogonal to L, and so the angle at p_π in $\triangle bp_\pi c$ is the card angle θ.

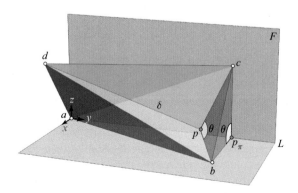

Figure 5.27 The dihedral angle, along dp, $\delta = \theta$. Here $\theta = 90°$.

(2) The lengths $|bp|$ and $|bp_\pi|$ are equal: $|bp| = |bp_\pi|$. This is because bp is an edge of cardstock and so of fixed length—half the base edge of the creased $\triangle bcd$—and bp coincides with bp_π when $\theta = 180°$. See again Fig. 5.25.

(3) Similarly $|cp| = |cp_\pi|$, because we insisted that the tetrahedron T is symmetric about L.

(4) Therefore, the two triangles $\triangle bp_\pi c$ and $\triangle bpc$ are congruent, as all three sides match in length: They share the edge bc, and we just showed that the other two edges are equal in length. So the angle at p in $\triangle bpc$ is also θ.

(5) Both $\triangle dpb$ and $\triangle dpc$ have right angles at p, because dp is the altitude of $\triangle bcd$ when $\theta = 180°$. See again Fig. 5.25. Therefore, $\triangle bpc$ is orthogonal to the crease dp.

(6) So the angle θ at p in $\triangle bpc$ measures the dihedral angle along bp: $\delta = \theta$.

This completes the proof of Theorem 5.3.

The V-fold theorem (Theorem 2.1) is so similar that it would be pleasing to derive one from the other. Perhaps an enterprising reader will see how to unify the two proofs.

Exercise 5.9 Practice: Tetrahedron Dihedral

The tetrahedron in Fig. 5.28 has edge lengths $|ad| = 1$, $|ab| = \sqrt{2}$, and $|bc| = 2$, with right-triangle faces $\triangle abd$ and $\triangle acd$ with $90°$ at a. What is the dihedral angle δ along the edge bc, i.e., between $\triangle bcd$ and $\triangle bca$?

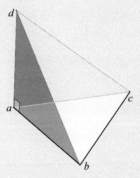

Figure 5.28 Tetrahedron with right angles at a.

5.7.2 Recursion

Because Theorem 5.3 establishes that the dihedral angle along the valley crease dp is θ, there is a sense that the triangles $\triangle dpb$ and $\triangle dpc$ form the back and front of an identically moving card: the card opening along L is "recapitulated" along the segment dp. This means that it is possible to build a type of recursion

Figure 5.29 Five regular tetrahedra all opening in concert.

Figure 5.30 A card that pops-up two regular tetrahedra. Animation available (O'Rourke, 2021).

into the pop-up, similar to the recursion we saw in Chapter 1, Fig. 1.19. For example, we can straddle another tetrahedron along dp. Indeed there is no reason to stop at two: Fig. 5.29 shows five tetrahedra each nested within the dp valley crease of the other.

In practice, there is a reason to stop at one: Even just adding one extra tetrahedron becomes a mechanical challenge. My crude attempt is shown in Fig. 5.30. Readers with better paper-engineering skills might create a more interesting pop-up card.

In this context we should note that Theorem 5.1 also permits a more straight-forward recursion: one cube on top of another!

Notes

For the Digital Michelangelo Project (Fig. 5.3), see (Levoy, Pulli, Curless, Rusinkiewicz, Koller, Pereira, Ginzton, Anderson, Davis, Ginsberg, et al., 2000) and `graphics.stanford.edu/papers/digmich_falletti/`. The original shopping bag proof is in (Balkcom, Demaine, and Demaine, 2004). Connelly's flexible polyhedron was announced in (Connelly, 1979), and the Bellows Theorem was settled in (Connelly, Sabitov, and Walz, 1997). Slits were explored in (Abel, Connelly, Demaine, Demaine, Hull, Lubiw, and Tachi, 2015). Flattening the Platonic solids was first achieved in (Itoh, Nara, and Vîlcu, 2012). Continuous flattening of any convex polyhedron is in (Abel, Demaine, Demaine, Itoh, Lubiw, Nara, and O'Rourke, 2014). The 180° cube pop-up design is from (Carter and Diaz, 1999). Rubber band pop-ups are described in (Johnson and Walser, 1997).

6

Algorithms for Pop-Up Design

6.1 Introduction

In this chapter we explore algorithms for constructing pop-up designs. Enthusiasts have developed software for pop-ups since the 1980s, but only recently have there emerged algorithms that achieve a measure of generality—but, as we will see, not yet full generality.

A pop-up algorithm takes as input some description of the desired pop-up structure, and it outputs templates and mountain/valley crease assignments that, when assembled and pasted in, pop up the structure as the card opens. An *algorithm* is a theoretical description of how the conversion from input to output is effected. Within its range of operation, the algorithm should always "work." Establishing this requires a "proof of correctness." Not all algorithms are practical, because they are either too complicated to implement in code or too slow in their execution, or both. Although there is now software all over the Web for designing a wide variety of pop-ups, they all have various limitations. They are practical but do not reach the generality that is the gold standard in mathematics.

Some of the most successful software apply to what is known as *origamic architecture*. This is a type of "pure" pop-up not dissimilar from "pure" origami, which insists on folding from a single piece of paper with no accoutrements, our focus in Chapters 1–3. Origamic architecture is created by cutting a single piece of cardstock, with no taped/glued additional pieces and no strings or rubber bands, but possibly with "windows" removed. Moreover, as the name suggests, the goal is a static structure at either card angle $\theta = 90°$ (the usual) or $\theta = 180°$. The dynamics of the structure upon card opening is irrelevant, the opposite of the pop-up spinner we studied in Chapter 4. Even though origamic architecture is a highly constrained model, pop-up artists have achieved stunning results, with no help from computers. We've seen several examples in Chapter 1; other impressive pop-ups are shown in Figs. 6.1. and 6.2.

And now with cutting machines allowing intricate templates (the template for Fig. 3.15 was machine-cut), the possibilities have expanded.

The structure popped up in origamic architecture is technically a *monotone surface*. Looking at a $\theta = 90°$ structure toward F, your line of sight intersects the structure at most once (and not at all if looking through a window). Here *mono* reflects the hitting-once aspect. The algorithms we will describe apply to non-monotonic surfaces.

A recent algorithmic advance, developed by a group of nine researchers, will be the focus of this chapter. Their three algorithms achieve generality in a sense explained shortly. However, the algorithms are progressively more complicated, perhaps impractical, and not yet fully implemented in software. Instead of attempting to explain the formidable details, I opt to describe the first algorithm in detail, from which it should be at least plausible that the second and third algorithms, which I only sketch, achieve their claimed goals.

First we need to explain a bit of terminology.

6.2   Orthogonal Polygons and Polyhedra

In Chapter 5 we explored a few special polyhedra—the Platonic solids—and explained what are more general polyhedra, both convex and nonconvex. In some sense, the ultimate pop-up goal would be to pop up an arbitrary polyhedron. Here *arbitrary* means "without any constraints," for us, nonconvex polyhedra with unlimited shape variation, although at the cost of perhaps having many vertices, edges, and faces. If an algorithm could be designed to pop up an arbitrary polyhedron, then it could, for example, pop up the head of David (Fig. 5.3) or the iconic Stanford Bunny shown in Fig. 6.3. How cool would that be!

Alas, this level of generality has not be reached. What has been achieved is pop-ups for any *orthogonal polyhedron*, which is still quite an impressive achievement. In an orthogonal polyhedron, all faces are parallel to two orthogonal planes, in our case, B and F at $\theta = 90°$. Another view is that each pair of faces that share an edge form a dihedral angle there of $90°$ or $270°$. Fig. 6.4 shows an example. This is the third algorithm that is too complicated to present in any detail. Instead we will describe a precursor algorithm that can pop up any extruded orthogonal polygon. So now let us define these terms.

As we saw in Chapter 5, a *polygon* is a non-self-intersecting closed path in the plane formed of straight-line segments, its *edges*, which meet at *vertices*. An *orthogonal polygon* is a polygon whose edges are parallel to two orthogonal lines, for example, the x- and y-axes. Equivalently, adjacent edges of a polygon meet at either $90°$, a *convex* vertex, or $270°$, a *reflex* vertex. Fig. 6.5 shows an example.

Figure 6.1 Rialto Bridge, Venice (Bianchini, Siliakus, and Aysta, 2009, p. 57). [©Random House. Used by permission of Potter Craft, an imprint of Random House, a division of Penguin Random House LLC. All rights reserved. Design and construction by Maria Victoria Garrido Bianchini.]

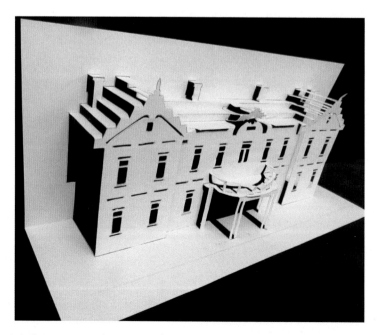

Figure 6.2 Origamic architecture. [Construction by Granville Hobson. Used by permission of the artist.]

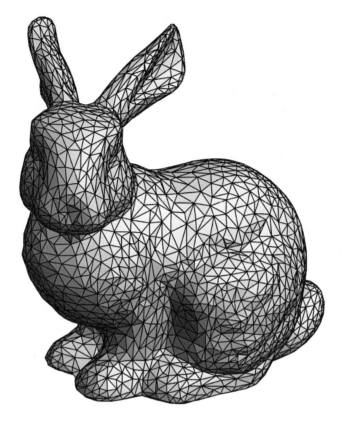

Figure 6.3 The Stanford Bunny: a polyhedron of 2503 vertices. [Model from the Stanford 3D Scanning Repository.]

Exercise 6.1 Understanding: Orthogonal Polygon

(1) If an orthogonal polygon P has r reflex vertices, how many convex vertices c must it have? Work out a conjecture.

(2) Prove your conjecture by reasoning about the angle turn at convex and reflex vertices.

Exercise 6.2 Challenge: Orthogonal Polygon Induction

Prove your answer to Exercise 6.1 by induction on r. (See Box 6.2.)

To base a pop-up structure on an orthogonal polygon, we *extrude* the polygon parallel to centerline L, as illustrated in Fig. 6.6. Admittedly, this does not necessarily make an interesting front view, but it is a step toward the more

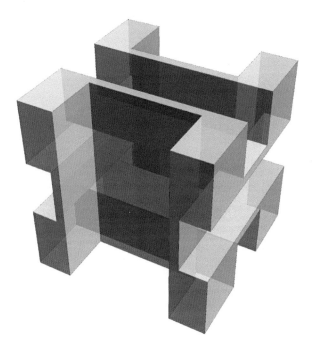

Figure 6.4 The octoplex orthogonal polyhedron. [Image by Emmely Rogers. For significance, see (Michael, 2009).]

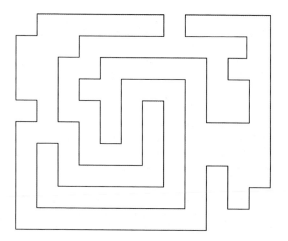

Figure 6.5 An orthogonal polygon.

interesting orthogonal polyhedra. Note that, in contrast to origamic architecture, the extrusion produces a non-monotone surface, in that rays from the front might pass through several layers.

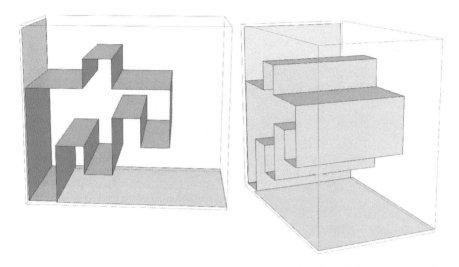

Figure 6.6 Extruded orthogonal polygon: side view (left) and front view (right).

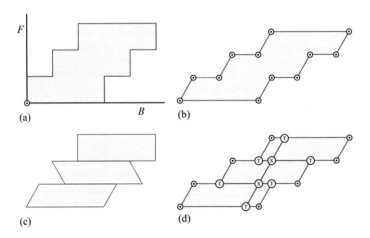

Figure 6.7 (a) Card fully opened. (b) All vertex joints. (c) Parallelograms. (d) Braced linkage.

6.3   Algorithm 1: Orthogonal Polygons

We have already used polygonal chains in Chapter 4: a sequence of fixed-length segments joined at universal joints, and forbidden to self-cross. The segments are often called *bars* to emphasize their fixed length. Here we will need a more general *linkage*, which extends the notion of a polygonal chain to a bar-and-joint linkage allowing more complex structures.

Fig. 6.7(a) shows an orthogonal polygon P fully opened at $\theta = 90°$. Ideally we would like to treat every vertex as a universal joint, so that P collapses as

θ is reduced toward 0°, as indicated in (b). However, each parallelogram can "shear" independently, as shown in (c). A solution is to brace P with additional struts and joints, as shown in (d). We now explain these ideas more formally.

6.3.1 Joints

Our goal is to arrange a linkage inside any given orthogonal polygon P so that it behaves just as in Fig. 6.7(a,b). This behavior can be described as a one degree-of-freedom (1-dof) motion that

(1) maintains the lengths of all edges,

(2) keeps all horizontal edges of P parallel and all vertical edges of P parallel,

(3) ensures the angles θ (convex) or $180° - \theta$ (reflex) at all the vertices of P.

The single degree of freedom is the card angle θ, with $\theta = 90°$ fully opened and $\theta = 0°$ fully closed. We call such a motion a *linkage-shear*; see Box 6.1.

Box 6.1 Shear

An example of a *shear* transformation is illustrated in Fig. 6.8. The y-coordinates of the vertices are unchanged, while the x-coordinates all increase by the same amount. But this means that the vertical edges increase in length. In a *linkage-shear*, all edges are bars whose lengths do not change.

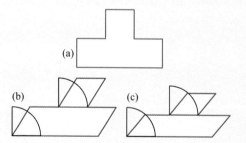

Figure 6.8 (a) Polygon. (b) Shear. (c) Linkage-shear.

Exercise 6.3 Practice: One Degree of Freedom

If all four vertices of a rectangle R are universal joints and the edges fixed-length bars, then, with its bottom edge fixed, R has just one degree of freedom. Does the L-shaped polygon in Fig. 6.9, similarly with all six vertices universal joints, also have just 1-dof?

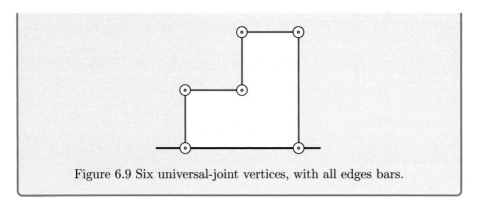

Figure 6.9 Six universal-joint vertices, with all edges bars.

The linkage will use three different type of joints, distinguished by their *degree*—the number of incident bars (see Fig. 6.10):

(1) degree-2 universal joints,

(3) degree-3 *T*-joints, or *flap* joints,

(3) degree-4 *X*-joints, or *slice-form* joints.

Universal joints are implemented by just a mountain or valley crease in the cardstock, which can bend forward and backward the full 360° range. *T*-joints are formed by attaching one piece of cardstock at an un-creased point of another. Such a "flap" has a 180° range.

An *X*-joint has two rigid bars through a point that form an *X*-shape. The full range of angles is possible, but the two bars remain straight through the joint. This can be realized by what is known as a *slice-form*, illustrated in Fig. 6.11. A thin, half-width slot is cut in two pieces of cardstock, then the two pieces are joined by opposing the slots, as in (c).

Returning to Fig. 6.7(d), it should be intuitively clear that the braced linkage should behave as desired. We now describe the algorithm for bracing an arbitrary orthogonal polygon, and then prove the correctness of the algorithm, i.e., prove that it always works. The proof will use induction; see Box 6.2.

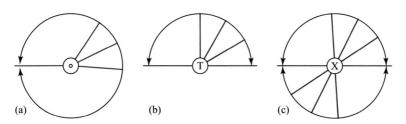

Figure 6.10 (a) Universal joint. (b) *T*-joint. (c) *X*-joint.

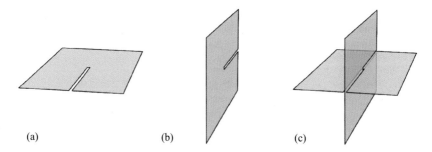

Figure 6.11 A slice form simulating an X-joint: (a) + (b) → (c).

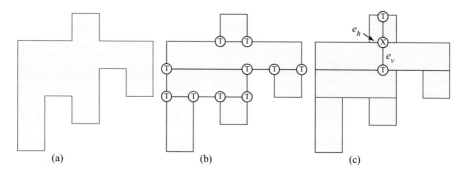

Figure 6.12 (a) P. (b) Addition of horizontal bars and T-joints. (c) Addition of a vertical brace and X-joint.

Box 6.2 Induction

Induction is a proof technique that can be used to establish that some claim is true for all numbers $n = 1, 2, 3, \ldots$. It is akin to climbing a ladder: if you know how to move from any one rung to the next, and you know how to reach the first rung, then you can climb to any rung, no matter how high. To reach the first rung, we only need to prove that the result holds for $n = 1$, the *base* of the induction. Moving from one rung to the next requires proving that if the theorem holds for $n - 1$ (you've reached that rung), then the theorem holds for n, where n is an arbitrary natural number. Then the theorem must be true for all n, "by induction" as they say: from $n = 1$, we can reach $n = 2$, and from there we can reach $n = 3$, and so on.

6.3.2 Algorithm 1

Let P be the orthogonal polygon in its fully opened, $\theta = 90°$ position. The algorithm consists of two stages, illustrated in Fig. 6.12.

(1) Extend each horizontal edge left and right internal to P with horizontal bars, partitioning P into rectangles. Where these bars touch the boundary of P, form a T-joint. See Fig. 6.12(b).

(2) For each pair of rectangles that share at least a portion of an extended horizontal edge e_h, and do not share a vertical edge (and are not already braced), brace them with a vertical bar e_v that has an X-joint at the intersection of e_h and e_v and T-joints at either endpoint of e_v. See Fig. 6.12(c). This brace is designed to prevent the independent linkage-shearing we saw in Fig. 6.7(c).

All other vertices of P are universal joints. Concerning the bracing struts, note that, for example, the three leftmost stacked rectangles in Fig 6.12(b) share a vertical edge and so do not need bracing. And always, a rectangle incident to the centerline L is effectively braced by the card front F.

Exercise 6.4 Understanding: Bracing

Apply the algorithm to create the linkage for the polygon in Fig. 6.13.

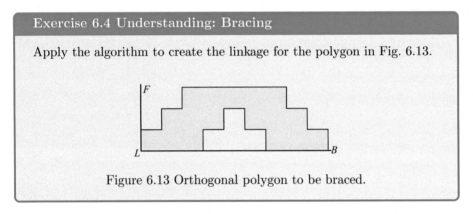

Figure 6.13 Orthogonal polygon to be braced.

6.3.3 Proof of Correctness

As mentioned, algorithms are intended to always work perfectly, under all inputs meeting the input criteria of the algorithm. Programs that only work sometimes using heuristics, or only produce approximate answers, are not technically "algorithms." Algorithms demand a "proof of correctness": a proof that when the inputs are legitimate, the algorithm will provably produce the desired output. In our case, the input is an orthogonal polygon P within the $B - F$ centerline card angle θ, and the output is a linkage defined inside P such that P will in fact collapse as a 1-dof mechanism to $\theta = 0$, or, equivalently, will pop up as θ increases from closed $\theta = 0°$ to $\theta = 90°$.

Box 6.3 Dual Graph and BFS

A graph consists of nodes and arcs connecting the nodes. The *dual graph* of the rectangular partition of an orthogonal polygon has a node for each rectangle and an arc between rectangles that share a line segment. See

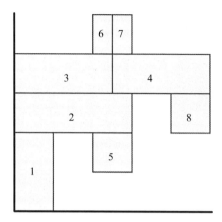

Figure 6.14 Numbering of rectangle cells.

Fig. 6.15(a). The ordering used in the proof is what is known as a *breadth-first search* (BFS) ordering of the graph nodes, a traversal that orders the nodes as shown in Fig. 6.15(b).

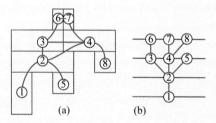

Figure 6.15 (a) Dual graph of an orthogonal polygon. (b) BFS ordering of the dual graph.

We now argue that, after the horizontal and vertical bars are added to P following the algorithm, the only motion of the resulting linkage is a linkage-shear: a 1-dof motion that maintains all edge lengths and slants all vertical edges by θ.

View the braced and partitioned polygon as consisting of a number of rectangular cells. For simplicity, assume one cell, C_1, is incident to the card centerline as in Fig. 6.14. (We skip the case where no cell is incident to the centerline, and just claim it can be handled similarly.) Number the rectangular cells starting with C_1 and then fanning out from there, as illustrated in Fig. 6.14. Say that two cells are *adjacent* if they share (a portion of) either an inserted horizontal bar or a vertical brace—any red segment in the figures. See Box 6.3. Now we argue by induction.

Cell C_1 may have horizontal bar inserts along its top edge and could have a vertical bracing bar as its right vertical edge. Regardless, the four edges of

C_1 are rigid bars, and so its only motion is to linkage-shear the rectangle to a parallelogram with lower-left corner angle θ. This settles the base of the induction.

For the general induction step, suppose that the linkage composed of cells $C_1, C_2, \ldots, C_k$ has the desired 1-dof linkage-shear motion. Consider adding in cell C_{k+1}, which by the numbering scheme is adjacent to at least one earlier cell, and perhaps more than one. For example, in Fig. 6.14, let $k = 3$. Cell $C_{k+1} = C_4$ is adjacent to both C_3 and C_2.

Let C_i be a cell adjacent to C_{k+1}, with $1 \le i \le k$. Consider two cases:

(1) Either a portion of one of C_{k+1}'s vertical edges is adjacently shared with C_i, or C_{k+1} and C_i share a vertical edge. For example, $C_{k+1} = C_4$ and $C_i = C_3$ are adjacent across a shared inserted vertical brace, whereas $C_{k+1} = C_5$ and $C_i = C_2$ share a common (right) vertical edge. Then, because that edge of C_i must slant by θ by the induction hypothesis, C_{k+1} is forced to slant the same.

(2) C_{k+1} and C_i are adjacent across (a portion of) a horizontal bar (and the previous case does not hold). For example, $C_{k+1} = C_6$ and $C_i = C_3$ fall into this case. Then the algorithm would have inserted a vertical brace and an X-joint connecting the two cells. In the example, the brace is between $C_3 \cup C_6$ and $C_4 \cup C_7$. Then the slanting of C_i must be mirrored by identical slanting of C_{k+1}.

The induction hypothesis guarantees the claim up to cell C_k, and we've now shown extension to C_{k+1}. Therefore, we conclude that indeed the entire linkage has just one degree of freedom, with each cell linkage-shearing to a parallelogram with lower-left corner angle θ, as in Fig. 6.16.

Although this algorithm is rarely practical, it has the advantage of being universal: every orthogonal polygon can be so realized. This universality is built upon in the next two algorithms.

6.4   Algorithm 2: General Polygons

More challenging and more interesting, the same authors generalized the algorithm to work for any polygon, similarly extruded parallel to the card centerline. An example is shown in Fig. 6.17. Very roughly, the algorithm proceeds as follows. First, a ray is defined from the centerline corner of the polygon P to every vertex of P. This partitions the interior of P into a number of quadrilateral and triangle cells. The addition of several more bars achieves a linkage that has a shear motion, but that shear is only one among many possible collapsing motions. The remainder of the construction inserts various additional supports and braces forming "gadgets" that restrict the flexibility of cells so that the linkage has just one degree of freedom, the desired linkage-shear motion, and avoids self-crossing throughout the motion.

Avoiding self-crossing is perhaps the most delicate part of the argument. Just to give some sense of the types of lemmas needed, we mention one here

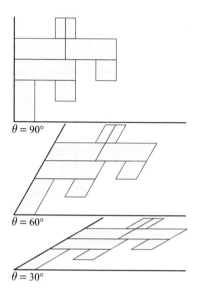

Figure 6.16 Shears at several values of θ. Animation available (O'Rourke, 2021).

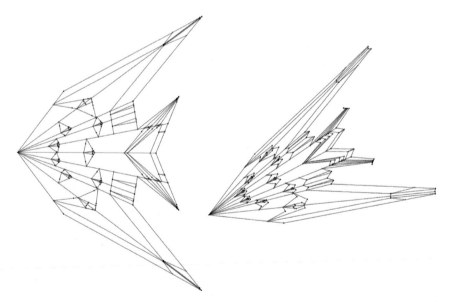

Figure 6.17 Fish polygon, where the apex at left is the card centerline: fully opened, a bit beyond 90° (left) and partially closed (right). Animation available (O'Rourke, 2021). [Snapshot from erikdemaine.org/papers/. Used by permission of Erik Demaine.]

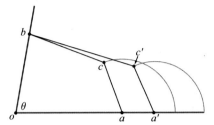

Figure 6.18 The chain acb never crosses $a'c'b$ as θ is decreased to $0°$. Animation available (O'Rourke, 2021).

without proof. Consider in Fig. 6.18 the two 4-link chains $oacb$ and $oa'c'b$ nested in a "V" formed by rays from o through a and through b. The V is hinged at o, whose angle is θ. The lemma is that, if joints c and c' are both perimeter midpoints of their 4-link chains, in the sense that the lengths of the two paths from o to c and from o to c' are equal,

$$|oa| + |ac| = |ob| + |bc|,$$
$$|oa'| + |a'c'| = |ob| + |bc'|,$$

then the 2-link chains acb and $a'c'b$ do not cross as θ is varied: they remain nested througout. Note that the circle arcs on which c and c' travel do in fact cross, so the claim is not obvious.

Several such lemmas are needed to prove that Algorithm 2 works for any polygon.

> **Exercise 6.5 Practice: Torus**
>
> Notice that the polyhedron in Fig. 6.19 is topologically a torus. What is the fewest number of vertices needed to construct an orthogonal polyhedral torus?

> **Exercise 6.6 Practice: Lattice Polyhedron**
>
> Define a *lattice polyhedron* as one whose vertices have integer coordinates, i.e., in $\mathbb{Z}^3$. What is the maximum number of vertices of a lattice polyhedron within $[0,3]^3$, i.e., whose coordinate values are in $\{0,1,2,3\}$?

6.5   Algorithm 3: Orthogonal Polyhedra

As is evident from Fig. 6.16, the area of the 2D polygon P decreases from its fully open position at $\theta = 90°$ to zero area at its fully collapsed state at $\theta = 0°$. One might imagine mimicking the 2D algorithm for a 3D polyhedron pop-up, where in place of rigid bars we have rigid polygonal faces, and in place of joints

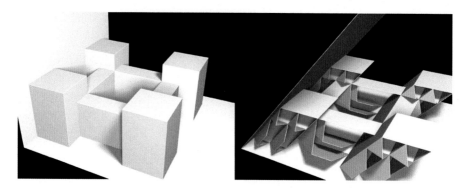

Figure 6.19 Castle model. Animation available (O'Rourke, 2021). [Snapshot from erikdemaine.org/papers/. Used by permission of Erik Demaine.]

shared by two bars, we have hinges along each edge shared by two faces. This would then be an instance of rigid origami, as discussed in Chapter 5.

As one might imagine, the algorithm for creating a pop-up of an arbitrary 3D orthogonal polyhedron is not straightforward. Instead of a linkage, the algorithm constructs a "scaffold" that partitions the interior into 3D cells, initially (at $\theta = 90°$) rectangular boxes. The intersection of this scaffold with a plane orthogonal to the centerline L is a 2D orthogonal polygon, an observation that is exploited in the algorithm. As before, the challenge is to design the scaffold to ensure that the motion of one cell is constrained by its adjacent cells to a single degree-of-freedom motion, and that pieces of the construction do not cross during card opening. This requires creases that we saw used in the cube pop-up in Chapter 5, Fig. 5.17. The authors call this a *draped scaffold*.

Their achievement is best appreciated through the animation prepared by the authors, two snapshots of which are shown in Fig. 6.19. Note how the rectangular side panels are creased at 45° and detached from two edges of the polyhedron so they can flex outward as θ varies, the same dynamic we saw in the cube pop-up.

The conclusion of Algorithm 3 is that the algorithm constructs a structure that pops-up <u>any</u> orthogonal polyhedron:

Theorem 6.1 Pop-Up Orthogonal Polyhedra

The draped scaffold of an orthogonal polyhedron P constitutes a pop-up for P.

The authors conclude their paper with this:

Open Problem 6.1 Algorithm for General Polyhedra

"The most obvious open question is whether there is a way to construct 3D pop-ups for general polyhedra."

 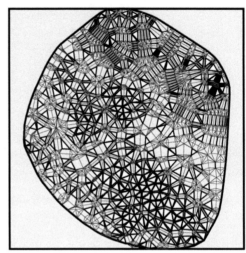

Figure 6.20 Origami Stanford Bunny (left) and its crease pattern (right) (Demaine and Tachi, 2017, Fig. 1). [Reprinted by permission of the authors.]

Recalling how complex might be a general polyhedron (Fig. 6.3), it is hardly surprising that this problem remains unsolved. Nevertheless, who would have predicted that the Stanford Bunny could be origami-folded from a single piece of paper? And yet a recent algorithm showed that any polyhedron could be so folded: see Fig. 6.20.

Finally, lest the reader worry that algorithm advances will drain the artistry out of pop-up design, consider these three remarks:

(1) The algorithms are not practical, and, as we will see in the next chapter, there is a sense in which true practicality may be forever out of reach.

(2) Algorithms may allow intricate constructions that would be difficult to achieve by hand, in the same way that laser cutters have capabilities beyond scissors and scapel knives. Artists can achieve more with better tools.

(3) So far, algorithms have concentrated on constructions whose goal is a static popped-up structure. I am unaware of algorithms that design for the dynamic effects that we saw, for example, in Chapters 2 and 4.

Notes

The algorithms in this chapter are all from (Abel, Demaine, Demaine, Eisenstat, Lubiw, Schulz, Souvaine, Viglietta, and Winslow, 2013). Masahiro Chatani, a pioneer in using computers to help design pop-ups, named the *origamic architecture* type of pop-up (Ando, Yamahata, Masumi, and Chatani, 2001). Impressive algorithms have been developed for designing a given origamic architecture structure. See, for example, (Mitani and Suzuki, 2004) and (Li, Shen, Huang, Ju, and Hu, 2010).

7

Pop-Up Design is Hard

We close with a brief discussion of something every pop-up designer knows intuitively: pop-up design is hard! Technically the result is that determining whether or not a particular pop-up design permits the card to open or to close is $\mathcal{NP}$-hard. We first explain the meaning of this "$\mathcal{NP}$-hard" term and then hint—and only hint—at the proof for pop-ups.

7.1 $\mathcal{NP}$-Hard Problems

Computer scientists have classified problems according to their "computational complexity" into two broad bins: those in the class $\mathcal{P}$, which are considered tractable, and all others, which are intractable, or $\mathcal{NP}$-hard. Here *tractable* means "easily solved," and *intractable* means "no one knows how to solve large instances"—the famous $\mathcal{P} = \mathcal{NP}$? problem. (And among the intractable problems, there are several finer classifications of the degree of intractability, which we will ignore in this sketch.)

Here is an example. Suppose that you have a set of natural numbers S, and the problem is to compute the median of the numbers. For example, perhaps this is your S:

$$S = \{11, 23, 7, 30, 14, 2, 3, 6, 5, 15, 32\} .$$

The median is 11: five numbers are below the median $(2, 3, 5, 6, 7)$ and five above $(14, 15, 23, 30, 32)$. This problem is in $\mathcal{P}$. "$\mathcal{P}$" stands for *Polynomial*, and in this case, the problem can be solved in what is called *linear time* with respect to the size n of S. This means that, as n grows, the time to find the median grows as a linear function (roughly, a multiple) of n. So this problem is considered tractable, even "easy."

An example of an intractable problem is SET PARTITION: given a set like S, can it be partitioned into two sets S_1 and S_2 such that the sum of the numbers in S_1 is equal to the sum of the numbers in S_2? In our example, the answer is "yes": for this partition, each sums to 74:

$$S_1 = \{2, 6, 11, 23, 32\} , \quad S_2 = \{3, 5, 7, 14, 15, 30\} .$$

However, if the number 2 in S is replaced by 9, forming a new set S', there is no equal-sum partition of S'. If n is the size of S, the only method known to find such a partition is to essentially try all possible partitions, which grows as the function 2^n. In the example, S has 11 elements, so that requires checking $2^{11} = 2048$ partitions—and that's how I determined there is no solution to the S' version. Various shortcuts and efficiencies are known, but still the time complexity grows as an exponential in n, not a polynomial. Checking 2^{11} partitions is easy, but if S contains, say, $n = 100$ numbers, then with $2^{100} > 10^{30}$, a brute-force search for a partition is quite infeasible even on the fastest computers.

Exercise 7.1 Practice: Set Partition

Consider $S = \{1, 2, 3, 4, 5, 6, 7, 8\}$. Can S be partitioned into S_1 and S_2 so that $\Sigma S_1 = \Sigma S_2$?

Although one might think that "$\mathcal{NP}$" stands for "non-polynomial," in fact it stands for *Non-derministically Polynomial*. It would take us far afield to explain this adequately, so we'll just be content with the informal $\mathcal{P}$ = tractable versus $\mathcal{NP}$-hard = intractable.

Many problems have been established as lying in one of these two classes. For example, almost all interesting games, when generalized for an arbitrarily large parameter n, are $\mathcal{NP}$-hard, including generalized Chess and Go, and even Mario and Pokémon! If a game were in $\mathcal{P}$, it would be too easy to be interesting.

7.2   Proving a Problem is $\mathcal{NP}$-Hard

The primary technique to proving that a problem X is $\mathcal{NP}$-hard is showing that, if one could solve X quickly (in *polynomial-time*), then one could solve some other problem known to be intractable quickly. This relies on a collection of known $\mathcal{NP}$-hard problems that can be "reduced" to problem X. To prove that it is intractable to determine if a particular pop-up structure can collapse to a closed position, a variant of a known-intractable problem is used: 3-SAT, standing for "3-variable satisfiability." This is a logic problem involving Boolean variables that can be either TRUE or FALSE, T or F. An example instance of 3-SAT is this:

$$(x_1 \text{ or } x_3 \text{ or } x_4)$$
$$\text{and} \quad ((\text{not } x_1) \text{ or } (\text{not } x_2) \text{ or } x_4)$$
$$\text{and} \quad (x_2 \text{ or } x_1 \text{ or } (\text{not } x_3)).$$

Each variable x_i or (not x_i) is called a *literal*. The "3" in 3-SAT indicates that each clause has three literals connected by "or"s. The problem is to find TRUE/FALSE assignments to the four variables that "satisfy" the expression, satisfy in the sense that the expression evaluates to TRUE. For this simple example, these assignments render the expression TRUE:

$$x_1 = T, \quad x_2 = F, \quad x_3 = F, \quad x_4 = F,$$

whereas the assignments $x_1 = x_3 = x_4 = F$ leave the first clause FALSE and so the whole expression FALSE (because it is a conjunction of clauses). Again, no one knows an algorithm for 3-SAT essentially superior than trying all T/F assignments, which takes time growing at the rate of 2^n for n variables.

Exercise 7.2 Practice: 3-SAT

Is this instance of 2-SAT satisfiable?

$$(x_1 \text{ or } x_2)$$
$$\text{and } ((\text{not } x_1) \text{ or } x_2)$$
$$\text{and } (x_1 \text{ or } (\text{not } x_2))$$
$$\text{and } ((\text{not } x_1) \text{ or } (\text{not } x_2))$$

7.3 Reducing 3-SAT to Pop-Up Design

This explanation will be quite incomplete. In particular, the proof for the intractability of collapsing uses a variant of 3-SAT called NAE3-SAT whose details I will ignore. The interested reader should consult the Notes for the original paper.

The overall structure of the proof is as follows. We are given a 3-SAT instance, with a certain number of variables and particular clauses as described previously. We then construct, following a blueprint of instructions, a pop-up mechanism between the front and back faces of a card that can only collapse closed, without adding new creases, if there is a satisfying assignment to the conjunction of the given clauses. So if we had a fast algorithm for deciding collapsibility, we would have a fast algorithm for 3-SAT. But it was proved in 1971 that 3-SAT is intractable, so collapsibility must be intractable also: $\mathcal{NP}$-hard.

Constructions to mimic 3-SAT are Rube Goldberg–like structures consisting of collections of interconnected *gadgets* that mimic aspects of the logic of 3-SAT in the structure—in our case, a pop-up construction. Here we opt to just show a variable gadget, whose role is to "set" a particular variable to TRUE or FALSE. Each variable gadget then needs to be connected to all the clause gadgets in which that variable occurs.

The variable gadget is illustrated in Fig. 7.1. The gadget consists of a rectangle with a rectangular hole, and various creases that can be either mountain or valley creases, or uncreased. Imagine a gadget fully opened flat, so that during card closure it must bend along the creases to compactify. (In this model, it is not permissible to add new creases.) There are several patterns of mountain/valley crease assignments that permit it to compactify, two of which are highlighted in the figure. One can see that a left tab protrudes in (b) and a right tab protrudes in (d). These two fundamentally different foldings are treated as TRUE and FALSE, respectively. There are two other foldings of this gadget that need to be inhibited by mechanisms (more gadgets) so that there are just two truth values, a typical detail needed in such constructions.

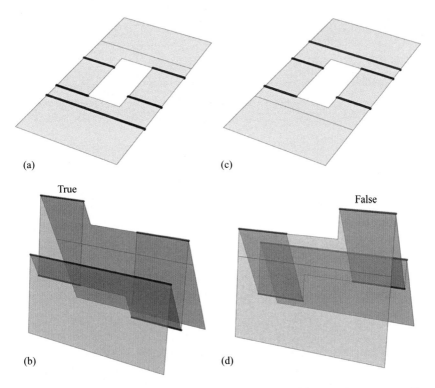

Figure 7.1 A variable gadget. Red: mountain crease. Blue: valley crease. Gray: flat crease.

Next a clause gadget needs to be designed so that it can compactify only if one of its three literals evaluates to TRUE (and, in the NAE3-SAT variant used here, also one literal must evaluate to FALSE). Without attempting to describe the clause gadget, we'll just mention that a "wire"—a strip of cardstock—can be attached to the TRUE or FALSE tab of the variable gadget, and from there connected to the clause gadget, conveying the variable setting to the clause. The design is such that the clause can collapse only if the wires tug in the correct distance amounts, which depends on the variable truth assignments.

Finally, all the gadgets must be hooked together to faithfully represent the logical structure of the given 3-SAT instance. The end result is that the card can close and compactify the whole structure if, and only if, there is a choice of truth assignments to the variables—which map into a choice of mountain/valley assignments in the variable gadgets—that render the 3-SAT instance TRUE. Thus, in this limited and technical sense, deciding whether a particular pop-up construction at $\theta = 90°$ will "work" in the sense of allowing the card to close to $\theta = 0°$, is $\mathcal{NP}$-hard. Some constructions simply cannot collapse flat, and deciding which cannot is intractable.

Standing back from the formidable technical details, the conclusion is that there is enough combinatorial complexity in mountain/valley assignments to a pre-creased structure to render intractable whether or not it can function as a pop-up design. However, returning to the algorithms discussed in Chapter 6, this intractability result does not imply that the hope for a resolution of Open Problem 6.1—an algorithm for general polyhedra—is in vain. Even $\mathcal{NP}$-hard problems often can be solved by non-polynomial algorithms if the combinatorial controlling parameter n is not too large. We mentioned that generalized Chess and generalized Go are intractable, but "generalized" here means variants on boards $n \times n$ for arbitrarily large n. For fixed-size boards, 8×8 for Chess and 19×19 for Go, computer algorithms have matched or exceeded the top grandmasters. Similarly, we may hope that some day there will be a computer-designed Stanford Bunny pop-up card (Fig. 6.3).

Notes

The proof that pop-up design is $\mathcal{NP}$-hard is in (Uehara and Teramoto, 2009). That many Nintendo games are $\mathcal{NP}$-hard is established in (Aloupis, Demaine, Guo, and Viglietta, 2015). Many games are what is known as $\mathcal{PSPACE}$-hard, the next level of difficulty beyond $\mathcal{NP}$-hard. See (Hearn and Demaine, 2006) for a comprehensive study of the complexity of combinatorial games.

8

Solutions to Exercises

Chapter 1 Exercises

Exercise 1.1 Solution: Within Card Profile

The required condition is $w + h \leq W$.

Exercise 1.2 Solution: Arcs Intersection

The two circles are

$$x^2 + z^2 = h^2 ,$$
$$(x - w)^2 + z^2 = h^2 .$$

We can solve these two equations simultaneously by, for example, subtracting the first from the second, which leaves

$$(x - w)^2 = x^2 .$$

Expanding and canceling terms leads to $x = \frac{w}{2}$. One hardly needs this algebra to see that the two congruent circles cross at the midpoint of the separation of their centers!

Substituting this value of x into $x^2 + z^2 = h^2$ leads to $z = \pm \frac{1}{2}\sqrt{4h^2 - w^2}$, with the $\pm$ indicating that there are two solutions, one above the card back B and one equally far below. So the crossing point p in space, assuming $y = \frac{W}{2}$, is

$$p = \left(\frac{w}{2}, \frac{W}{2}, \frac{1}{2}\sqrt{4h^2 - w^2} \right) .$$

Exercise 1.3 Solution: Parallel Unequal

When the card is closed flat, the paper between the cuts is angled at 90°, parallel to L. So it turns 90° from the fully open card to the closed card.

Exercise 1.4 Solution: Three Parallel Cuts

The two strips between the cuts attempt to pass through one another. If you force the card closed, the 45° valley creases are ignored, and new creases are introduced to effectively make the cuts the same length.

Exercise 1.5 Solution: Slanted Parallel

The parallelogram between the cuts develops a mountain fold halfway between the slanted valley folds, and with the same slant. At $\theta = 90°$ the result is a slanted table.

Exercise 1.6 Solution: Tallest Letter

Examining Fig. 1.14, the maximum letter height is reached at $h + 2w = W$. So if w is chosen very small, say, $w = \varepsilon$, then $h = W - 2\varepsilon$ can be nearly W. Of course then the letter only pops out by ε. (The symbol ε generally represents a small, positive number.)

Exercise 1.7 Solution: Half-Cylinder

The pop-up requires parallel cuts terminating on semicircles: see Fig. 8.1.

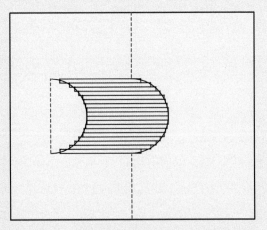

Figure 8.1 Red: Mountain creases. Blue: Valley creases.

Chapter 2 Exercises

Exercise 2.1 Solution: Parametric Ellipse

The parametric equations are

$$x(t) = a \cos t \,,$$
$$y(t) = b \sin t \,,$$

with t in $[0, 360°]$. For example, when $t = 0°$, $(x, y) = (a, 0)$; when $t = 90°$, $(x, y) = (0, b)$.

Exercise 2.2 Solution: Intersection of Two Circles

Let the two centers be $a = (0, 0)$ and $b = (1, 0)$. The points of intersection $p = (x, y)$ must satisfy $|pa| = |pb| = 1$. This yields the two equations (using the squared lengths)

$$x^2 + y^2 = 1^2 \,,$$
$$(x - 1)^2 + y^2 = 1^2 \,.$$

Subtracting the two equations to eliminate y yields $x^2 - (x - 1)^2 = 0$, which solves to $x = \frac{1}{2}$. Substituting this into either equation leads to $y^2 = \frac{3}{4}$, so the two points are $p = (\frac{1}{2}, \pm\frac{\sqrt{3}}{2})$.

Exercise 2.3 Solution: Three Spheres, One Point

If $|ab| = 2$, then the two sphere $S_a \cap S_b$ just touch at a single point p. Then any point c with $|pc| = 1$ leads to $S_a \cap S_b \cap S_c = p$.

Another configuration is this. Let $|ab| < 2$ so that $S_a \cap S_b = C$ is a circle C. Then one can place c so that S_c is just tangent to C and $S_c \cap C$ is a single point.

Exercise 2.4 Solution: $\delta = \theta$ when $\alpha \neq 45°$

The argument does not depend on α. Here are the six steps of the argument reviewed (succinctly):

1, 2 : The angle of the triangle $\triangle bcp_\pi$ at p_π is still θ.

3, 4 : Both $\triangle abp$ and $\triangle acp$ are still right triangles at p, so δ is $\angle bpc$ at p.

5, 6 : The triangles $\triangle bcp_\pi$ and $\triangle bcp$ are still congruent, and so $\delta = \theta$.

The argument does depend on $|bp| = |cp|$ to guarantee congruency of $\triangle bcp_\pi$ and $\triangle bcp$.

Exercise 2.5 Solution: Sphere ∩ Circle

Although Fig. 2.11 shows only the single point of intersection p, there is a second intersection below B, with a negative z-coordinate.

Exercise 2.6 Solution: Full p-arc

See Fig. 8.2. The blue triangle $\triangle ap_\pi c_\pi$ and the pink triangle $\triangle ap_0 c_0$ are congruent. The length $|ap_\pi| = |ap_0|$ because both are positions of the same crease ap. $|ac_\pi| = |ac_0|$ because c_π reflects across L to reach c_0. The angle at a in both triangles is $2\alpha = 90°$. Therefore, by the side-angle-side theorem, the two triangles are congruent, and so their third sides are equal: $|c_\pi p_\pi| = |c_0 p_0|$.

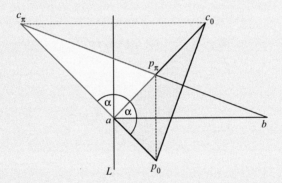

Figure 8.2 The blue and pink triangles are congruent.

Exercise 2.7 Solution: V-fold angle $\alpha \neq 45°$

The total rotation is approximately 2α, only approximately for the reasons mentioned in the text.

Chapter 3 Exercises

Exercise 3.1 Solution: Rim at $\theta = 90°$

When $\theta = 90°$, the centermost rib is vertical from the card back B to the rim, and horizontal from the card front F to the rim. So p is the center point of the rim. Its distance from L is $\sqrt{2}$ because the centermost rib constitutes a radius of the unit circle C, and so p's distance to L is the hypotenuse of a 1-by-1 isosceles right triangle.

Exercise 3.2 Solution: Reflection

See Fig. 8.3:

- $(0,0) \rightarrow (1,1)$
- $(\frac{1}{2},0) \rightarrow (1,\frac{1}{2})$
- $(\frac{1}{2},\frac{3}{2}) \rightarrow (-\frac{1}{2},\frac{1}{2})$

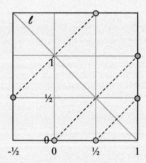

Figure 8.3 Reflections in line ℓ.

Exercise 3.3 Solution: Verify r^3

Indeed $c_y = r^3$. Because $s = \frac{1}{2}$, $r = \sqrt{1 - \frac{1}{4}} = \frac{\sqrt{3}}{2} \approx 0.87$, and $c_y = r^3 = (\frac{\sqrt{3}}{2})^3 = \frac{3\sqrt{3}}{8} \approx 0.65$.

Exercise 3.4 Solution: Epicycloid

Such curves are called *epicycloids*. If the two circles have the same radius, the traced curve is a *cardioid* and touches C just once, at the start/end point $(1,0)$. The cardiod is heart-shaped; the name comes from the Greek *kardia* meaning "heart." Animations are widely available on the Web.

Exercise 3.5 Solution: 3D Parametric Equations

- $\theta = 0°$: Because $\tan 0 = 0$, $z = 0$, which is correct. With $\cos 0 = 1$, both denominators of the x and y equations reduce to 2. So then we have

$$x(s,0) = s(3 - 2s^2),$$
$$y(s,0) = 2(1 - s^2)^{3/2}.$$

These both match Eq. 3.6.

- $\theta = 180°$: Because $\cos 90° = 0$, $y(s, 180°) = 0$, and so also $z(s, 180°) = 0$. (Actually, z leads to $\infty \times 0$, but the "race" to 0 driven by $\cos 90°$ is faster than the race to ∞ driven by $\tan 90° = \infty$, a fact that needs calculus.) Using $\cos 180° = -1$ leads, after simplification, to $x(s, 180°) = s$. So indeed we just have the diameter of C along x: $(s, 0, 0)$.

Exercise 3.6 Solution: Parabola Equation

$$y = 1 - \frac{x^2}{4}.$$

For example, at $x = 0$, $y = 1$, the vertex $v = (0, 1)$; at $x = 2$, $y = 0$, where the radius-2 circle C meets the x-axis.

Chapter 4 Exercises

Exercise 4.1 Solution: Planar 3-Link Maxspan

There are only two distinct planar configurations of a 3-link $90°$ chain: either two left turns at the joints, or a left and right turn; see Fig. 8.4. The first case leads to a span of $\sqrt{2^2 + 2^2} = \sqrt{8} \approx 2.8$, while the second staircase configuration achieves the maxspan of $\sqrt{4^2 + 2^2} = \sqrt{20} \approx 4.5$.

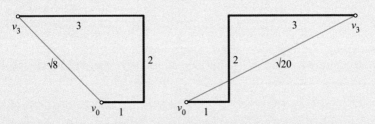

Figure 8.4 The two planar configurations of a $(1, 2, 3)$ $90°$-chain.

Exercise 4.2 Solution: 3-Link Maxspan in 3D

No, the span cannot be increased by rotating the third link out of the plane. Fig. 8.5 shows why.

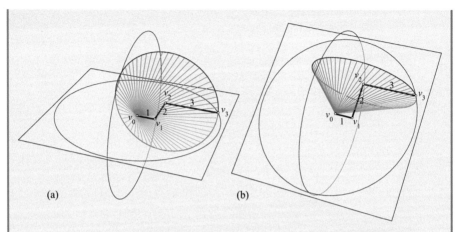

(a)   (b)

Figure 8.5 Two views of rotating v_3 out of the plane of $\{v_0, v_1, v_2\}$.

If the first two links are kept fixed and the third is rotated through its full freedom of motion, v_3, the tip of the third link, traces out a circle of radius 3 centered on v_2, with the disk of the circle perpendicular to the v_1v_2 middle link. The question then becomes: are any of the points on this circle (blue in the figure) farther away from v_0 than is the planar span of $\sqrt{20}$? Those distances from v_0 are drawn green in the figure. That they are all shorter than $\sqrt{20}$ becomes more plausible if we imagine a sphere of radius $\sqrt{20}$ centered on v_0, indicated by the two red great circles in the figure. The blue circle of possible v_3 positions lies entirely inside the $\sqrt{20}$ sphere and just touches it at the one point that corresponds to the maxspan, achieved by the staircase configuration.

In fact it is a theorem (due to Nadia Benbernou (Benbernou, 2006)) that the maxspan of any 3-link fixed-angle chain is achieved in a planar configuration.

Exercise 4.3 Solution: Odd-n maxspan

Fix one endpoint at $(0,0)$, and let the first link be horizontal, to $(1,0)$. Then the other endpoint has coordinates

$$\left(\left\lceil \frac{n}{2} \right\rceil, \left\lfloor \frac{n}{2} \right\rfloor\right) = \left(\frac{n+1}{2}, \frac{n-1}{2}\right),$$

where the symbols $\lfloor\ \rfloor$ and $\lceil\ \rceil$ indicate rounding down and up, respectively. So the end-to-end distance is

$$\sqrt{\left(\frac{n+1}{2}\right)^2 + \left(\frac{n-1}{2}\right)^2} = \frac{\sqrt{n^2+1}}{\sqrt{2}}.$$

Note that this last expression is close to the even-n maxspan: $n/\sqrt{2}$.

Exercise 4.4 Solution: Minimal Spinner

(1) Is the radius r of C fixed? *No.* Circle C is centered on the midpoint of $a'b'$ and is the intersection of unit-radius spheres centered on a' and b': $C = S_{a'} \cap S_{b'}$. Because a' approaches b' as the card closes, the intersection of the spheres changes. At

$$\theta = 180°, \ 90°, \ 0° \ ,$$

the radius r of C is

$$\sqrt{2}/2, \ \sqrt{3}/2, \ 1 \ ,$$

respectively.

(2) How far does c travel angularly? Point c traverses a 90° arc in M, half the travel of c'. See the animation (O'Rourke, 2021).

Chapter 5 Exercises

Exercise 5.1 Solution: Truncated Icosahedron

The hexagon angle is 120° and the pentagon angle is 108°. So $2 \times 120° + 108° = 348°$, just 12° shy of 360°.

Exercise 5.2 Solution: Nonconvex Platonic Solid

See Fig. 8.6.

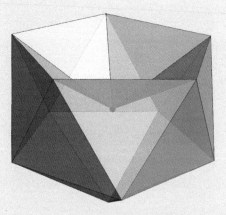

Figure 8.6 Dented icosahedron.

Exercise 5.3 Solution: Cube + Octahedron

The polyhedron is just barely nonconvex: see Fig. 8.7. Note that 360° surrounds the cube vertices.

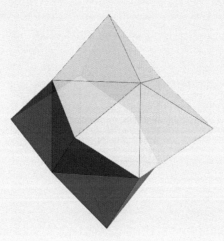

Figure 8.7 Octahedron pyramids on the faces of a cube.

If one shortens the apex-to-square-base distance until the triangles are right isosceles, then one obtains the convex *tetrakis hexahedron*.

Exercise 5.4 Solution: Rigid Cube

- The cube remains rigid with one face removed.
- It remains rigid with two adjacent faces removed.
- Without both the top and bottom faces, the resulting chain of four squares has the same flexibility as a rhombus.

Exercise 5.5 Solution: Flattened Cube Area

Yes. One can reduce the area by changing the top crease to mountain and the side creases to valley. This achieves an area of $1\frac{1}{2}$.

Exercise 5.6 Solution: Cube 90°

Vertex v_5 traverses a quarter-arc of a circle centered on v_1. Vertex v_5' also traverses a quarter-arc of a circle centered on v_1, with the two circles lying in perpendicular planes. So both travel a distance of $\pi/2$. This is perhaps clearest in the animation (O'Rourke, 2021).

Exercise 5.7 Solution: Two Triangle Dihedrals

The two triangles are both unit-edge-length equilateral triangles. So their altitudes have length $\sqrt{3}/2$. Draw the altitudes from e to the midpoint m of ab, and from f to m. Let 2α be the angle of the triangle $\triangle emf$ at m. Then $2\alpha = \delta_{ab}$. We have

$$\sin\alpha = \frac{(1/2)|ef|}{\sqrt{3}/2}\ ,$$

and now solving for α yields $\delta_{ab} = 2\sin^{-1}(|ef|/\sqrt{3})$.

Exercise 5.8 Solution: Not Great Circles

Let V be the hypothetical vertical plane through points d_1, b, and p, the north pole of S_b; V must also include the $\theta = 0°$ position of d_1. Consider the three triangles incident to b: $\triangle bd_1e, \triangle bea, \triangle baf$. They form a $180°$ angle at b initially. So f_0 is also on V; see Fig. 8.8.

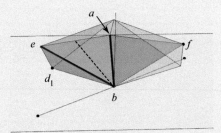

Figure 8.8 Highlighted: $\delta_{ba} = \delta_{be}$.

We know that $\delta_{ba} = \delta_{be}$ from Theorem 5.2. This implies that the configuration of those three triangles is symmetric about the line through b and the midpoint $m = (a + e)/2$, the altitude foot on ae. Because $d_1 \in V$ by the hypothesis, $f \in V$ also. Yet we know that f does not travel on a great circle, but rather on a vertical plane containing e and f.

Interestingly, $\triangle d_1bf$ <u>is</u> vertical at all times, but the vertical plane containing that triangle is not fixed: instead it rotates a bit.

Exercise 5.9 Solution: Tetrahedron Dihedral

The dihedral angle is $45°$. In Fig. 8.9, p is the midpoint of bc, so $|pa| = 1$ and $|pd| = \sqrt{2}$. So $\delta = \cos^{-1}(1/\sqrt{2}) = 45°$.

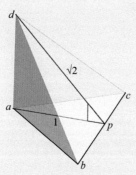

Figure 8.9 Dihedral angle: $\cos 45° = 1/\sqrt{2}$.

Chapter 6 Exercises

Exercise 6.1 Solution: Orthogonal Polygon

(1) $c = r + 4$.

(2) Imagine walking around the boundary of P, counterclockwise. At each convex vertex, you turn left by $90°$. At each reflex vertex, you turn right by $90°$. By the time you return to your starting point and starting orientation, you must have turned $360°$. So you must turn left four more times than right: $c = r + 4$.

Exercise 6.2 Solution: Orthogonal Polygon Induction

Let $n = c + r$ be the total number of vertices of P. *Base*: When $r = 0$, P is a rectangle, and $c = 4$. *General Step*: Select any reflex vertex a, and extend a vertical cut from a to a point b on the boundary of P. There are two situations: b is on the interior of an edge of P, or b is another reflex vertex. (It is not possible for b to be a convex vertex.) See Fig. 8.10.

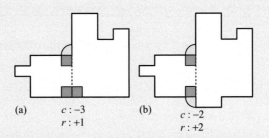

Figure 8.10 Over- and under-counts in c and in r.

By the induction hypothesis, the vertex counts for the two subpolygons are $c_1 = r_1 + 4$ and $c_2 = r_2 + 4$. One can see that in the first situation, (a) in the figure, $c_1 + c_2$ over-counts c by 3 convex vertices (green), and $r_1 + r_2$ under-counts r by 1 reflex vertex (yellow). So we have

$$
\begin{aligned}
c &= c_1 + c_2 - 3 \\
&= (r_1 + 4) + (r_2 + 4) - 3 \\
&= r_1 + r_2 + 5 \\
&= (r - 1) + 5 \\
&= r + 4 .
\end{aligned}
$$

The (b) situation follows from a similar calculation.

Exercise 6.3 Solution: One Degree of Freedom

No, the L-polygon has more than one dof: see Fig. 8.11. In fact, it has three degrees of freedom. Calculating the dof of a linkage is a topic in engineering mechanics.

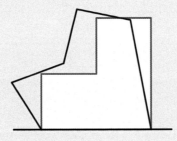

Figure 8.11 The L-polygon has three degrees of freedom.

Exercise 6.4 Solution: Bracing

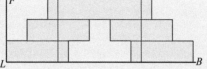

Figure 8.12 The braced orthogonal polygon.

Exercise 6.5 Solution: Torus

An orthogonal polyhedral torus needs at least 16 vertices: 8 for the outer shell, and 8 for the hole. (The same question for general (non-orthogonal) polyhedra is quite difficult. There the answer is 7, the Császár polyhedron built from 14 triangles.)

Exercise 6.6 Solution: Lattice Polyhedron

The polyhedron in Fig. 8.13 has 32 vertices.

Figure 8.13 Polycube built from seven cubes: six surrounding a central cube.

Chapter 7 Exercises

Exercise 7.1 Solution: Set Partition

Yes:

$$S_1 = \{1, 2, 3, 5, 7\} , \ S_2 = \{4, 6, 8\} ,$$
$$\Sigma S_1 = \Sigma S_2 = 18 .$$

Exercise 7.2 Solution: 3-SAT

The expression is not satisfiable:

- If $x_1 = T$, then the second clause forces $x_2 = T$. But then the fourth clause is F.

- If $x_1 = F$, then the first clause forces $x_2 = T$. But then the third clause is F.

List of Symbols

Symbol	Meaning	First Page(s)
$a, b, c, \ldots$	Specific points, edges, or lengths	1
$\alpha, \ldots$	Angles (labeled with Greek letters)	1
F, B, L	Card Front, Back, and centerLine, respectively	1
θ	Card angle: dihedral angle between F and B, along L; in $[0°, 180°]$	1
δ	Dihedral angle between two faces lying in two planes	2
x, y, z	Coordinates	2
$|ab|$	Length of line segment ab	4
W, H	Card width and height, respectively; both F and B are $W \times H$ rectangles	6
p_0, p_π	Position of point p when $\theta = 0°$ and when $\theta = 180°$, respectively	15
$\triangle abc$	Triangle with vertices a, b, and c	15
M	Medial plane, halfway between F and B	16
1-dof	one degree-of-freedom	16, 91
$\pm$	Indicates two expressions, one preceded by $+$ and the other by $-$	18, 107
$\approx$	Approximately equal	19
$\in$	Is a member of; for example, $\theta \in [0°, 180°]$ (not the same as epsilon, ε)	19
$[a, b]$	Interval on the real line from a to b, including both a and b	19
$\cos^{-1}$, arccos	Inverse cosine; similarly, $\sin^{-1}$ or arcsin is the inverse sine	19
$\cap$	Intersection of sets; $S_b \cap S_c$ ($\cup$ is used for union)	20
$p_\perp$	Perpendicular projection of point p	22
$a', b', c', \ldots$	Points closely related to $a, b, c, \ldots$; pronounced, e.g., a-prime	35
$v_1, v_2, v_3, \ldots$	Vertices	52

$\angle abc$	Angle at b between line segments ba and bc	75
P	Generally reserved for polygon or polyhedron	90
$\mathbb{Z}^3$	The 3D integer lattice. $\mathbb{Z}$ represents the integers	98
$\mathcal{P}, \mathcal{NP}$	Complexity classes: Polynomial and Nondeterministically Polynomial	101
ε	Small, positive real number (not the same as set membership, $\in$)	108
$\lfloor x \rfloor, \lceil x \rceil$	Floor and ceiling of x: round x down/up to the nearest integer	113
Σ	Sum	120

Bibliography

Abel, Z., R. Connelly, E. D. Demaine, M. L. Demaine, T. C. Hull, A. Lubiw, and T. Tachi (2015). Rigid flattening of polyhedra with slits. In *Origami*[6]*: Proceedings of the Sixth International Meeting on Origami Science, Mathematics and Education*, pp. 109–117. American Mathematical Society.

Abel, Z., E. D. Demaine, M. L. Demaine, J.-i. Itoh, A. Lubiw, C. Nara, and J. O'Rourke (2014). Continuously flattening polyhedra using straight skeletons. In *Proceedings of the Thirtieth Annual Symposium on Computational Geometry*, pp. 396–405. Association for Computing Machinery.

Abel, Z., E. D. Demaine, M. L. Demaine, S. C. Eisenstat, A. Lubiw, A. Schulz, D. L. Souvaine, G. Viglietta, and A. Winslow (2013). Algorithms for designing pop-up cards. In *Proceedings of the 30th International Symposium on Theoretical Aspects of Computer Science*, Leibniz International Proceedings in Informatics Volume 20, pp. 269–280. Schloss Dagstuhl Publishing.

Aloupis, G., E. D. Demaine, A. Guo, and G. Viglietta (2015). Classic Nintendo games are (computationally) hard. *Theoretical Computer Science 586*, 135–160.

Ando, N., N. Yamahata, S. Masumi, and M. Chatani (2001). Shape grammar and form properties of architectural figures. *Journal for Geometry and Graphics 5*(1), 23–33.

Balkcom, D. J., E. D. Demaine, and M. L. Demaine (2004). Folding paper shopping bags. In *Proceedings of the 14th Fall Workshop on Computational Geometry*, pp. 14–15. Massachusetts Institute of Technology.

Benbernou, N. (2006). *Fixed-Angle Polygonal Chains: Locked Chains and the Maximum Span*. Undergraduate thesis, Smith College.

Benbernou, N. (2011). *Geometric Algorithms for Reconfigurable Structures*. Ph.D. thesis, Massachusetts Institute of Technology.

Benbernou, N. and J. O'Rourke (2006). On the maximum span of fixed-angle chains. In *Proceedings of the 18th Canadian Conference on Computational Geometry*, pp. 93–96. CCCG.

Bianchini, M. V. G., I. Siliakus, and J. Aysta (2009). *The Paper Architect: Fold-It-Yourself Buildings and Structures with 20 Ready-to-Use Templates.* Potter Craft.

Carter, D. A. and J. Diaz (1999). *The Elements of Pop-Up: A Pop-Up Book for Aspiring Paper Engineers.* Little Simon.

Connelly, R. (1979). How to build a flexible polyhedral surface. In *Geometric Topology*, pp. 675–683. Elsevier.

Connelly, R., I. Sabitov, and A. Walz (1997). The bellows conjecture. *Beiträge zur Algebra und Geometrie 38*, 1–10.

Davis, D. (2020, October). How to cut and fold a spinner. YouTube video: `www.youtube.com/watch?v=YkHa24IIhis&ab_channel=DianaDavis`.

del Rosario Ruiz, Jr., C. (2015). *Automated Paper Pop-Up Design: Approximating Shape and Motion.* Ph.D. thesis, National University of Singapore.

Demaine, E. D. and J. O'Rourke (2007). *Geometric Folding Algorithms: Linkages, Origami, Polyhedra.* Cambridge University Press.

Demaine, E. D. and T. Tachi (2017). Origamizer: A practical algorithm for folding any polyhedron. In *Proceedings of the 33rd International Symposium on Computational Geometry*, pp. 34:1–34:15. Dagstuhl-Leibniz-Zentrum für Informatik.

Hearn, R. A. and E. D. Demaine (2006). *Games, Puzzles, and Computation.* A K Peters/CRC Press.

Hull, T. C. (2020). *Origametry: Mathematical Methods in Paper Folding.* Cambridge University Press.

Itoh, J.-i., C. Nara, and C. Vîlcu (2012). Continuous flattening of convex polyhedra. In *Computational Geometry*, Lecture Notes in Computer Science, Volume 7579, pp. 85–97. Springer.

Jackson, P. (1993). *The Pop-Up Book: Step-by-Step Instructions for Creating Over 100 Original Paper Projects.* Henry Holt and Co.

Jakus, S. and J. O'Rourke (2012). From pop-up cards to coffee-cup caustics: The knight's visor. arXiv:1206.1312.

Johnson, S. and H. Walser (1997). Pop-up polyhedra. *The Mathematical Gazette 81*(492), 364–380.

Levoy, M., K. Pulli, B. Curless, S. Rusinkiewicz, D. Koller, L. Pereira, M. Ginzton, S. Anderson, J. Davis, J. Ginsberg, et al. (2000). The digital Michelangelo project: 3D scanning of large statues. In *Proceedings of the 27th Annual Conference on Computer Graphics and Interactive Techniques*, pp. 131–144. Association for Computing Machinery.

Li, X.-Y., C.-H. Shen, S.-S. Huang, T. Ju, and S.-M. Hu (2010). Popup: Automatic paper architectures from 3D models. *ACM Transactions on Graphics 29*(4), 111.

Michael, T. S. (2009). *How to Guard an Art Gallery and Other Discrete Mathematical Adventures*. Johns Hopkins University Press.

Mitani, J. and H. Suzuki (2004). Computer aided design for origamic architecture models with polygonal representation. In *Proceedings Computer Graphics International*, pp. 93–99. IEEE.

O'Rourke, J. (2011). *How to Fold It: The Mathematics of Linkages, Origami, and Polyhedra*. Cambridge University Press.

O'Rourke, J. (2021). Pop-up animated GIFs. `cs.smith.edu/~jorourke/PopUps/`.

Uehara, R. and S. Teramoto (2009). Computational complexity of a pop-up book. In *Origami⁴: Fourth International Meeting of Origami Science, Mathematics, and Education*, pp. 307–316. A K Peters/CRC Press.

Valette, S., J. M. Chassery, and R. Prost (2008). Generic remeshing of 3D triangular meshes with metric-dependent discrete Voronoi diagrams. *IEEE Transactions on Visualization and Computer Graphics 14* (2), 369–381.

Wells, D. (1991). *The Penguin Directory of Curious and Interesting Geometry*. Penguin.

Zirbel, S. A., R. J. Lang, M. W. Thomson, D. A. Sigel, P. E. Walkemeyer, B. P. Trease, S. P. Magleby, and L. L. Howell (2013). Accommodating thickness in origami-based deployable arrays. *Journal of Mechanical Design 135* (11), 111005.

Index

Printed in the United States
by Baker & Taylor Publisher Services